Advances in Information Systems and Business Engineering

Edited by
U. Baumöl, Hagen, Germany
J. vom Brocke, Vaduz, Liechtenstein
R. Jung, St. Gallen, Switzerland

The series presents current research findings based on diverse research methods at the interface between information systems research, computer science, and management science. The publications in this series aim at practical concepts, models, methods, but also theories which address the role of information systems in the innovative design and sustainable development of organizations. Contributions are selected which on the one hand propose innovative approaches how modern information and communication technologies can enable new business models. On the other hand, contributions are eligible which present considerable improvements to existing solutions. The publications are characterized by a holistic approach. They account for the socio-technical nature of work-systems and suggest that the development and implementation of information systems need to consider the multi-faceted organizational context driven by people, tasks, and technology.

Edited by
Prof. Dr. Ulrike Baumöl
FernUniversität Hagen, Germany

Prof. Dr. Reinhard Jung,
Universität St. Gallen, Switzerland

Prof. Dr. Jan vom Brocke
Universität Liechtenstein, Fürstentum
Liechtenstein

Francesca Ricciardi

Innovation Processes in Business Networks

Managing Inter-Organizational Relationships for Innovational Excellence

Francesca Ricciardi
Torino, Italy

Printed with the friendly support of Fondazione CUOA.

CENTRO UNIVERSITARIO
DI ORGANIZZAZIONE AZIENDALE

Business School dal 1957

ISBN 978-3-658-03438-2 ISBN 978-3-658-03439-9 (eBook)
DOI 10.1007/978-3-658-03439-9

The Deutsche Nationalbibliothek lists this publication in the Deutsche Nationalbibliografie; detailed bib-
liographic data are available in the Internet at http://dnb.d-nb.de.

Library of Congress Control Number: 2013946557

Springer Gabler
© Springer Fachmedien Wiesbaden 2014

This work is subject to copyright. All rights are reserved by the Publisher, whether the whole or part of
the material is concerned, specifically the rights of translation, reprinting, reuse of illustrations, recitation,
broadcasting, reproduction on microfilms or in any other physical way, and transmission or information
storage and retrieval, electronic adaptation, computer software, or by similar or dissimilar methodology
now known or hereafter developed. Exempted from this legal reservation are brief excerpts in connec-
tion with reviews or scholarly analysis or material supplied specifically for the purpose of being entered
and executed on a computer system, for exclusive use by the purchaser of the work. Duplication of this
publication or parts thereof is permitted only under the provisions of the Copyright Law of the Publisher's
location, in its current version, and permission for use must always be obtained from Springer. Permissions
for use may be obtained through RightsLink at the Copyright Clearance Center. Violations are liable to
prosecution under the respective Copyright Law.
The use of general descriptive names, registered names, trademarks, service marks, etc. in this publication
does not imply, even in the absence of a specific statement, that such names are exempt from the relevant
protective laws and regulations and therefore free for general use.
While the advice and information in this book are believed to be true and accurate at the date of publication, neither the authors nor the editors nor the publisher can accept any legal responsibility for any errors
or omissions that may be made. The publisher makes no warranty, express or implied, with respect to the
material contained herein.

Printed on acid-free paper

Springer Gabler is a brand of Springer DE.
Springer DE is part of Springer Science+Business Media.
www.springer-gabler.de

Foreword

Innovation has become a crucial challenge in today's management processes.

As a consequence, management scholars feel more and more driven to investigate how innovation capabilities can be created, encouraged and fostered in organizations.

Many of these capabilities are influenced by the way the organization conceives, designs and manages its network of business relationships. Therefore, many factors that may be decisive to innovation, although external to the organization's boundaries, are actually impacted by the organization's choices (or inertia). For example, constructive cooperation with software suppliers is often one of the most important success factors for information systems innovation.

Management literature has been working intensely on this fascinating issue in the last years, and has developed some key constructs that have become classical explanatory variables for business network success, such as, for example, Trust or Information Sharing.

But further research is needed in this field. In fact, there are many real-world situations in which it is the business network itself that seems to hinder successful and sustainable innovations, even if the traditional business network variables are ranking high.

As a matter of fact, most extant theories (e.g. the Resource Based View of business networks, discussed in influential works like those by Combs and Ketchen in the 1990s) are more interested in the "bright side" of business networking; thus, they see inter-organizational networks as substantially positive phenomena, provided they are managed with the right tools. But there are also theories (e.g. the theory of the Population Ecology of Organizations, developed by Hannan and Freeman since the 1970s) that are more interested, instead, in the "dark side" of long-term business relationships. If we compare theories of the former and of the latter type, we must admit that they tend to see business networks through very different constructs and explanatory patterns. Not surprisingly, these different theories make different predictions as for the impacts of networking activities on innovation-related performances.

When and why does a real-world business network fall into a certain theory and not into another? Under what conditions does the theory of the Population Ecology of Organizations, for example, become an appropriate model for understanding the specific business network we are looking at? What is the link, if there is any, between the many different theories proposing so different explanations and predictions on the same important phenomenon?

Management studies, in other words, are today challenged to build a unified theory of business networking, capable to explain both the cases in which business networks trigger dynamism and effective change, and the cases in which they promote opacity, conservatism, competitive blindness, or persistent commitment to ruinous innovation strategies.

This is the gap in theory this book concentrates on.

The contribution of this work consists in a novel theoretical approach to innovation processes in business networks. This approach seeks to leverage the potentialities of extant theories and to exploit the complementarities of different scholarly traditions, including management studies, evolutionary studies, game theories, and system theories.

Three original theoretical reflections are developed throughout the book: a theory of innovation, a theory of altruistic cooperation and fairness, and a theory of motivations for business networking. These three theories are linked together in the last Chapter, where a modular theory of innovation in business networking is presented.

Chapter 4 points out the contradictory outcomes of field research on creativity and innovation in organizational settings. Then, a quite original and challenging explanation of the dynamics of innovation is presented, that is consistent with the seemingly paradoxical results collected by the scholarly community so far.

The theory of cooperation and fairness presented in Chapter 3, on the other side, is strongly rooted in game theories and system theories, and sees inter-organizational altruism as a dynamic system of behaviors whose success is a key factor for business network strength.

Finally, the theory of motivations for business networking presented in Chapter 5 provides a possible explanation of the very different attitudes towards innovation that can be observed in different business networks.

In this book, innovation-related performances are measured not only in terms of competitive excellence, but also in terms of social excellence. In fact, this work aims to explain the ethical impacts of business networking, too: they may range from very positive to extremely negative, as happens, for example, in opaque lobbying.

Thanks to expert panel discussions, the theoretical concepts developed in this book are translated into twelve constructs, and the related measurement scales are developed. All the theory-building efforts of this work are eventually translated into testable models; as a consequence, they are easily usable for further research.

For this reason, not only does this book provide the reader with a detailed possible unified theory of innovation in networked business settings, but it also encourages and facilitates further scholarly work on issues that are perceived as highly relevant by the world of practice.

Prof. Cecilia Rossignoli
University of Verona, Italy

Acknowledgements

This book stems from a re-elaboration and re-writing of my doctoral thesis, discussed at the Catholic University of Milan, Italy, in 2013, and of some papers I have written during the last five years. I want to warmly thank the friends and colleagues with whom I have discussed the topics addressed in this book, and especially Chiara Cantù, Patrizia Lombardi, Isabella Maggioni and Elena Marcoz.

Marco De Marco has caringly followed my scholarly work since the beginning: he has been the first one to drive me to engage in scholarly research, and he has always supported me with his valuable advice.

Cecilia Rossignoli has strongly encouraged me to write this monograph. Her experience and extraordinary capabilities in coordinating and managing the research activities have really been a model to me.

Jan vom Brocke introduced me to the international community of researchers; his example has been very important to make me understand how scientific research should be designed and presented.

I also want to thank Barbara Imperatori, for encouraging me to address the complex issue of organizational creativity as an antecedent of innovation.

My tutor, Renato Fiocca, challenged me to investigate the possibility of altruistic cooperation in business settings, which was the core of my PhD thesis: I am really grateful for having been proposed such an inspiring topic.

The people I interviewed, and the expert panel members I have discussed my outcomes with, were essential to my work: I want to express my gratitude for their collaboration, even if, for confidentiality agreements, I cannot mention their names here.

Francesca Ricciardi

Table of Contents

XII

List of Figures

List of Tables

1. Introduction

This theory-building research focuses on how business networking influences the organization's performances, and more specifically on how business networking influences the organization's capabilities of achieving excellence through innovation.

In this study, an organization's Business Network is understood as the dynamic system of the organization's long-term business relationships.

According to this definition, not only is a Business Network the footprint of the organization's past relational experiences, but it is also the expression of the organization's ideas about the future: in fact, the relationship with a certain external subject is considered by an organization as a part of its own Business Network only to the extent that further interactions with that specific external subject are perceived by the organization as likely, needed and/or desirable.

Another corollary of this definition is that the list of the possible external subjects involved in an organization's Business Network includes not only other firms, but also all the other actors with which long-term business-oriented interactions may occur: for example, Public Administration (PA) bodies, professionals, consumers, associations, opinion makers, educational institutions, research centers, politic parties, local leaders, etc.

Business networks are, for example, well-established supplier-purchaser value chains, industrial districts, tourist destinations, integrated supply chains, joint ventures, lobbies, professional associations, business clubs, etc.

Consistently with the wide definition of Business Network adopted here, then, this study will take into consideration the whole range of possible business network interactions: in other words, this work will seek to identify models suitable not only for the classical B2B network relationships, which are typically at the core of most mainstream literature, but also for other relationships, such as, for example, G2B (Government to Business) or C2G (Citizens to Government, e.g. participatory initiatives).

In fact, the organization's relationships with different categories of external subjects strongly influence each other. For example, if the relationships with a PA body or with a political party prove very successful in providing the organization with valuable resources and protection, the organization may be less incentivized to invest, say, in a demanding B2B partnership aimed at new product development. In other words, in order to get the whole picture of the reasons and consequences of business networking choices, it is probably important to holistically describe the complete horizon of the organization's relational life.

Of course, this research purpose, although challenging and stimulating, implied some difficulties. In fact, it was going to include, in the same theoretical framework, even very different phenomena, such as, for example, lobbying activities, on the one side, and creative inter-organizational efforts for product co-design, on the other side.

In effect, studies on the different types of interactions in business networking activities have been usually conducted by different research communities so far, each adopting its own language and its specific goals.

The theoretical approaches taken into consideration for this study are nine, five of which are Organizational / Management Theories, and namely:

Organizational / Management Theories:

1. Theory of the Population Ecology Of Organizations (Organizational Ecology)
2. Resource Dependence Theory
3. Resource Based View / Relational Based View of the Firm
4. Collaborative Network Theory
5. Institutional Theory.

Views from other Disciplines:

1. Evolutionary Studies
2. Human Ethology
3. Adaptive Dynamics
4. Game Theories.

The literature analyses I conducted, more thoroughly described in the following Chapters, did not result in the discovery of any comprehensive model, taking into account all the different aspects of business networking synthesized above. More specifically, we lack theoretical tools for understanding those cases, recurring in the world of practice, where it is the Business Network itself that seems to hinder the organization's innovation capabilities and competitiveness, even if all the "traditional" variables (such as Trust, Cooperation, Number of Ties, Resource Flows, Attitude to Legality, etc.) are ranking high.

I then decided to take on this challenge, and to design this work as a theory-building research, aimed at giving a contribution to fill this gap.

After literature survey, I identified five key topics to explore, in order to frame my theoretical and empirical researches.

These are:

a. Understand the whole range of possible positive and negative impacts of business networking on performances.
b. Understand the role of cooperation processes in business networks.
c. Understand the dynamic processes of creativity and innovation and how they relate to business networking.
d. Understand how business networks are shaped by the actors' motivations for networking.
e. Understand how the factors above may interact with (IT enabled) inter-organizational processes and practices in creating excellence through business networking.

Each of these topics was analyzed through literature review and theoretical reflection; moreover, for all the topics interviews were conducted for ex-post model validation and fine-tuning, whilst topic "d" was also investigated through explorative, qualitative field research.

This work is theory-building oriented, and consistently each of the five investigation streams presented above resulted in the definition of specific constructs and scales, which were discussed for face validation with (i) three experts in business networking from a leading Italian bank, with (ii) a manager from an international leading consulting firm specialized in supply chain management, and (iii) with the general manager of a company that promotes tourism in Italy.

These researches allowed me not only to propose 12 scales to measure a significant range of key business networking features and consequences, but also to hypothesize a model connecting the 12 constructs in cause-effect relationships.

This book is then structured as in Table 1.1.

Table 1.1. Book Structure

Chapt.	Res. Method	Research Purpose	Theory Building Outcome
2	Expert Panel Validation	Understand the whole range of possible positive and negative impacts of business networking.	Constructs and Scales: *Competitive Excellence; Social Excellence; Innovational Success.*
3	Expert Panel Validation	Understand the role of altruism-enabled behaviors in inter-organizational interactions..	Construct and Scale: *Business Network Strength.*
4	Expert Panel Validation and Qualitative Explorative Research	Understand the dynamic processes of creativity and innovation and how they relate to business networking.	Constructs and Scales: *Strategic Support to Innovation; Organization's Innovation Capabilities.*
5	Expert Panel Validation	Understand how business networks are shaped by the actors' motivations for networking.	6 Constructs and Scales: *Motivations for Business Networking.*
6	Expert Panel Validation	Understand how business networking influences the organization's capabilities of achieving excellence through innovation (general Research Question)	A *Testable Model* including all the 12 Constructs above, and linking them through hypothesized cause-effect relationships.

2. Business Networking: Possible Positive and Negative Impacts on Innovation and Excellence

Abstract The purpose of this Chapter is to comprehensively examine a wide range of possible positive and negative impacts of Business Networking on the organizations' performances, and particularly on innovation-related performances. The final outcomes of this chapter are: (i) a framework, summarizing all the identified possible positive and negative impacts of business networking on the organization's performances, divided into five categories (Efficiency; Effectiveness; Innovation Capabilities; Social Role; Financial Strength); (ii) three Constructs and related Scales, designed for measuring the main innovation-related performances of organizations in empirical theory-testing research, and namely: Organization's Innovational Success; Organization's Competitive Excellence; Organization's Social Excellence.

2.1 Introduction

Literature reviews revealed that it is difficult to find exhaustive descriptions of all the possible impacts of business networking on performances: in other words, each disciplinary tradition tends to concentrate on only few specific possible (expected or unintended) consequences of business networking.

For example, the Industrial Network Approach belongs to B2B Marketing studies; scholars belonging to this research tradition usually investigate the mechanism through which cooperative Business Networking positively impacts on organizations' long-term market performances (Håkansson and Snehota, 1989), whilst negative impacts of business networking are only marginally studied (Ford et al., 2003). This approach is compatible with the Relational View of the Firm, that asserts that not only is an organization's network of relationships a means to access valuable external resources: the network of relationship should be understood as a most strategic resource *per se*. In fact, long-term interactions, if effectively managed, result in idiosyncratic environments that are unique and impossible to imitate: each individual innovation developed in a business network might be quickly imitable by competitors, but the specific innovational capabilities of an inter-organizational system cannot be replicated, and are then a powerful source of sustainable competitive advantage. This theory, in other words, states that the relational resources of an organization are an essential factor for building competition-effective innovation. On the other hand, the idiosyncratic nature of business networks implies only that each network's innovation capabilities are difficult to imitate: it does not imply that these capabilities are necessarily effective. Some publications highlight cases where idiosyncratic business networks result in over-embeddedness and inward-looking culture, i.e. in environments where the capabilities to *successfully* (and not only inimitably) innovate regrettably atrophy, instead of developing.

The Industrial Network Approach, on the other hand, can be considered as an important research community within the Collaborative Network Theory approach, which is considered as a key theory for explaining business networking behaviors, and which will be repeatedly referred to throughout this book.

Another interesting example may be Supply Chain Management studies. These studies focus on inter-firm process integration and on technologies for smoother resource/information flows; scholars belonging to this research tradition usually evaluate the performances of Business Networking in terms of information-based reactiveness, logistic efficiency and reduction of costs and risks (Chandra and Kumar, 2001). Some of these studies are dedicated not only to expected, but also to possible unintended consequences of supply chain management processes in business networks: negative impacts of business networking are ascribed to flawed processes mainly (Lee et al., 1997) and the role of Information Systems is decribed as growingly important to manage such complex interorganizational processes (Akyuz and Rehan, 2009). Supply Chain Management studies traditionally refer to classical theories on inter-organizational behaviors, such as the Transaction Costs Theory and the Resource Dependence Theory, which will be further analyzed throughout this book.

In order to understand also large-scale B2C (Business to Consumer) or G2C (Government or Public Administration Body to Citizen) networking, two scholarly traditions were considered for this study: Social Network studies, which concentrate on the socio-technical features of networking (Kadushin, 2012); and e-Participation studies, which concentrate on network-based and participatory management of common and public resources that may be vital for businesses (Sæbø et al., 2008), such as e.g. landscapes (for the tourism industry, for example) or transportation infrastructures (for the manufacturing sector, for example). These studies tend to be quite optimistic and to concentrate on positive outcomes; in this scholarly tradition, (business) networking, i.e. the system of network-based long-term interaction of public and private organizations with consumers/citizens, is expected to result in social improvements, such as Consumer Empowerment, Improved Civic Awareness, Improved Civic Engagement, Enhanced Civil Rights, Increased Territorial Value, Improved Territorial Services (Ricciardi and Lombardi, 2010).

In the Resource Based View of the Firm, inter-organizational interactions are seen as means to exploit the heterogeneity and complementarity of firms, so that each firm can access strategic resources beyond its own boundaries, in order to enhance competitive advantage (Combs and Ketchen 1999). According to this approach, business networking allows firms to share and/or exchange valuable, rare, inimitable and/or non-substitutable resources: possible negative impacts of business networking are hardly mentioned in these studies (Chen and Chen 2003).

Two streams of studies seem more strongly focused on the possible negative impacts of business networking: studies on lobbies and power groups, and studies based on the Theory of the Population Ecology of Organizations.

The former group of publications investigate the effects of favoritism, political patronage and lobbying on markets: these studies highlight the possible negative consequences of the opacity made possible by business networking, especially in those cases where strong personal interests, politics, financial powers and/or mob ties are involved (Zingales, 2009).

The Theory of the Population Ecology of Organizations, on the other side, predicts that the same inter-organizational interactions that select specialized, highly reliable organizations in a business settings, end up by making those organizations rigid and incapable to further change and evolve (Hannan and Freeman, 1989).

In a nutshell, each individual scholarly tradition, if taken in isolation, seems incapable of providing us with a comprehensive understanding of all the possible consequences of Business Networking on performances. On the other side, if considered as a whole, the sum of the different scholarly traditions provides us with a rich and satisfactory list of possible positive and negative impacts of business networking processes.

In the following part of this Chapter, the outcomes of this comparative literature review will be framed as follows:

- Possible impacts of Business Networking on the organization's efficiency;
- Possible impacts of Business Networking on the organization's effectiveness;
- Possible impacts of Business Networking on the organization's innovation capabilities;
- Possible impact of Business Networking on the organization's social role;
- Possible impacts of Business Networking on the organization's financial strength.

2.2 Possible Impacts of Business Networking on the Organization's Efficiency

The scholarly tradition that has more intensely concentrated on the expected efficiency enhancements driven by inter-organizational networking is the Supply Chain Management one.

In this stream of studies, long-term network interactions are expected to enhance control on critical resources (Hillman et al., 2009), and to reduce risks and costs, by making inter-organizational flows of information, things, money, decisions, etc., smoother and more rationally structured (Naylor et al., 1999). But some of these writings also highlight that such strong process integration (ideally resulting in an "extended enterprise") is very costly (Jagdev and Browne, 1998) and advantages may not compensate for the drawbacks, especially considering that higher complexity and difficulties in network governance may paradoxically result in system fragility and system crash (Melnyk, 2009).

Transaction Cost literature also highlights the beneficial decreases in transaction costs attainable thanks to long-term inter-organizational networking (Williamson, 1985).

2.3 Possible Impacts of Business Networking on the Organization's Effectiveness

Many scholarly writings insist on the improvements in the organization's effectiveness, that can be achieved thanks to Business Networking.

According to these writings, Business Networking allows access to valuable market understanding, and the joint resources of network partners can be usefully leveraged to achieve higher customer satisfaction (or citizen satisfaction, in case of Public Administration bodies), higher service quality, higher reactivity and flexibility (Ford et al., 2003). Moreover, some writings highlight that Business Networking, providing organizations with a more protected and predictable environment (Powell, 1996), makes more resources available for long-term investments and market effectiveness.

On the other hand, some writings highlight that the environment provided by Business Networking, i.e. a protected and predictable relational context of established, privileged relationships, may result in organizational rigidity and resistance to change, which is likely to result in organizational death (Hannan and Freeman, 1989). Moreover, the strongest the network bonds, the strongest is the possible amplification of certain local mistakes that may severely affect other partnering organizations' effectiveness (e.g. the so-called "bullwhip effect" in supply chains) and the contagion (McFarland et al., 2008) of possible failures and related reputation problems.

2.4 Possible Impacts of Business Networking on the Organization's Innovation Capabilities

The positive effects of Business Networking on the organization's innovation capabilities are thoroughly investigated by studies referring to the Collaborative Network theory, by studies referring to the Resource-Based View, and also by Social Network studies.

According to these studies, Business Networking allows access to valuable resources (information, competences, skills, patents, experiences, etc.) external to the individual organization's boundaries (Combs and Ketchen 1999): as a consequence, an innovation path can be taken, that would have been impossible to go through for the single organization alone. Moreover, if reciprocal trust between network partners is high (Ford et al., 2003), Business Networking allows rational division of labor, specialization, risk sharing, cultural exchange and cross-fertilization: all factors that are expected to positively influence the organization's innovation capabilities.

Also studies on Social Network are usually very confident of the innovative potential of the "collective intelligence" unfolding in social networking (Woolley et al., 2010). But in some cases, also this stream of studies expresses concern for the waste of time and resources that the very nature of network interactions may imply: some cases are reported, where networks perform poorly as for innovation capabilities, when compared to traditional, structured organizations (Sun et al., 2010).

On the other hand, also the Industrial network perspective sometimes mentions possible negative impacts of networking on innovation capabilities. Håkansson et al. (2009) describe some paradoxical aspects of business networks: each organization seeks to control its network, but control is destructive, since the more a single actor is able to exercise control over a network, the more the actor becomes the sole source of innovation. Moreover, the growing bonds between actors enable them, but simultaneously also constrain them.

The Institutional Theory explains innovation in business networks as a consequence of mimetic processes, in some cases triggered by coercive or normative forces (Meyer and Rowan, 1977). This means that decision-making as for innovation is influenced by institutional factors (such as, for example, the managers' prestige), which have little to do with the innovation's real importance and potentialities. This can lead the organization both towards constructive and destructive innovation-related decisions.

Table 2.1. Possible positive and negative impacts of Business Networking on the organization's performances

	Positive / Expected Impacts of Business Networking	Negative / Feared Impacts of Business Networking
On Efficiency	Rationalization of information and/or resource flows. Enhanced predictability and control. Lower transaction costs.	Uncontrolled costs of process integration. Conflicts, dispersive power games. Higher risks of system crashes due to higher complexity.
On Effectiveness	Better market understanding. Enhanced reactivity and flexibility, then service quality, then customer satisfaction.	Dependence-oriented culture. Laziness due to "network comfort". Responsibility "holes". (Inter-) organizational rigid lock-ins.
On Innovation Capabilities	Access to valuable resources allowing innovation. Higher specialization. Beneficial cultural exchanges and debates. Inter-organizational mimetic and coercive pressures may drive useful change.	Inward-looking network culture may result in innovational "blindness". Co-innovation is complex and often dispersive, may drift and result in wasted resources. Inter-organizational mimetic and coercive pressures may drive harmful change, or hinder useful change.
On Social Role	Attitude to social embeddedness. Care for Reputation. Enhanced social responsibility.	Over-exploitation of common and/or public goods. Opacity, protection of unfair privileges and of illegal behaviors. Fair competition hindered. Alternative initiatives hindered.
On Financial Strength	Better business credit scores. Enhanced business flows predictability. Reliability of partners' payments. Mutual financial help.	Financial fragility, due to the dependence on few key partners. The "Bow to the stronger partner's will" principle may jeopardize weaker partners in the long run.

2.5 Possible Impact of Business Networking on the Organization's Social Role

An interesting group of studies focuses on the social performances of Business Networks. On the one side, some authors (for example, those referring to Institutional Theory) say that Business Networking enhances the organization's attitude to cooperation, to social embeddedness and to reputation care (Meyer and Rowan, 1977): as a consequence, networked organizations are more likely to be engaged in socially responsible behaviors (Suchman, 1995).

On the other side, other authors highlight how Business Networks tend to evolve into power systems dedicated to over-exploit common and public goods (Ostrom and Walker, 2005), to cover up illegal behaviors, to protect unfair privileges, to hinder alternative initiatives, and to discourage civic engagement (Zingales, 2009): as a consequence, Business Networks can impact even very negatively on the social fabric.

2.6 Possible Impacts of Business Networking on the Organization's Financial Strength

After defining through Literature Review the aforementioned four possible Business Networking performance measures (organizational efficiency, effectiveness, innovation, social role), extracted from literature review, I discussed and fine-tuned them with the panel of experts enrolled in this research, i.e. three experts in Business Networking from a leading Italian bank, a manager from an international leading consulting firm specialized in supply chain management, and the general manager of a company that promotes tourism in Italy.

The three experts from the bank suggested to consider also a further performance measure, i.e. Financial Strength. In fact, they commented that one of the main expectations of organizations entering a Business Network is to have better business credit scores on the part of banks, which may take into consideration the enhanced value of an organization with well-established and promising business relationships. Moreover, long-term network partners often can rely on more predictable and reliable payment behaviors and, in some (although rare) cases, on direct financial help on the part of partners.

The manager of the consulting firm specialized in supply chain management, interviewed after this suggestion, agreed on that, but added that Business Networking may also result in a double-blade weapon as for the organization's financial strength. In fact, a strong financial dependency on few network partners may make the organization's financial situation dangerously fragile.

Finally, the general manager of the Italian local tourism promotion company, interviewed in turn, suggested further financial problems that may raise due to Business Networking: in case of opportunistic behaviors on the part of the stronger partners (as for discount requests or delayed payments, for example) some weaker actors of the network may feel forced to bow to the stronger partner's will, even if this results in financial problems.

As a consequence, also the organization's financial strength was considered as meaningful performance measurement for successful (or unsuccessful) Business Networking strategies, and included into the model. Table 2.1 summarizes the five areas of predicted impacts of business networking processes.

2.7 Measuring Business Network Impact

The classical performance data, such as profits or sales, may describe the possible impact of business networking on the organization in the present, but they do not capture the possible positive or negative impact of networking on the organization's capability of coping with the future, nor do they shed light on the possible positive or negative impact of the organization's networking strategies on the environment, i.e., for example, on the local community or on the territory.

For this reason, three different and complementary constructs were chosen, to provide a tool for a comprehensive evaluation of possible network impacts. These constructs are thought to complement the extant and well-established measures of financial performances and efficiency.

The first construct is "Organization's Innovational Success". This construct describes the extent to which the organization is capable of finding and developing new and successful solutions; this construct derives from the category "Innovation Capabilities" of Table 2.1 and is proposed here as a proxy of the organization's competition capability in prospect.

The second construct is "Organization's Competitive Excellence". This construct describes the extent to which the organization is strong today as for sales, market position and financial stability. This construct derives from the categories "Efficiency", "Effectiveness" and "Financial Strength" of Table 2.1, and is proposed here as a proxy of the organization's competition capability today.

The third construct is "Organization's Social Excellence". This construct derives from the category "Social Role" of Table 2.1 and describes the extent to which the organization can be considered as playing a positive role in the territory and in the social fabric it belongs to.

I developed three scales for these three constructs, on the basis of the theoretical work synthesized in Table 2.1, of the discussion with the expert involved in this research, and of the scales tested in Khanwalla and Mehta (2004) and Turker (2009). These three scales are presented in the following Paragraphs.

2.7.1 Organization's Innovational Success (IS)

8 Items Scale

Do you agree with the following statements? (1=strongly disagree, 5=strongly agree, 3=neutral).

IS-1
Our organization has an excellent image of being innovative, and our innovations are often a benchmark for competitors.
IS-2
An important part of our current revenues were derived from recent product / service innovation.
IS-3
Our organization has implemented a stream of successful innovations in business processes and/or organizational culture.
IS-4
Our organization has re-designed its competitive environment, for example by entering new markets or by modifying its supply network.
IS-5 (reverse item)
An important part of our organization's innovation projects has recently failed, because it proved incompatible with the status quo (e.g., with old work habits, with established powers, with our legacy information system, etc.)
IS-6 (reverse item)
An important part of our organization's innovation projects failed in the last years, because their costs proved too high.
IS-7 (reverse item)
Sometimes I fear that we are too specialized or too dependent from specific solutions or specific conditions, and that we would be hardly capable to adapt to changing conditions.
IS-8
Our organization has not made serious mistakes in choosing the technological tools for supporting innovations in the last years.

2.7.2 Organization's Social Excellence (SE)

10 Items Scale

Do you agree with the following statements? (1=strongly disagree, 5=strongly agree, 3=neutral).

SE-1
Our organization implements effective programs to minimize its negative impact on the environment.
SE-2
Our organization supports non-governmental organizations working in problematic areas and/or contributes to campaigns and projects that promote the well-being of the society.
SE-3
Our organization effectively contributes to improve the value and image of the territory (e.g., the city) it operates in.
SE-4
Our organization supports schools and/or universities and/or research centers, and/or funds scholarships.
SE-5
Our organization supports employees who want to acquire additional education.
SE-6
The managerial decisions related with the employees are usually fair in our organization.
SE-7
Our organization selects socially responsible business partners (e.g. suppliers) only.
SE-8
Our organization targets interaction fairness in its business relationships.
SE-9 (reverse item)
Our organization's image has been damaged by boycotts or scandals or legal actions in the last three years.
SE-10
Our organization complies with legal regulations completely and promptly.

2.7.3 Organization's Competitive Excellence (CE)

5 Items Scale

Do you agree with the following statements? (1=strongly disagree, 5=strongly agree, 3=neutral).

CE-1
Our organization's profitability has satisfied our investors/owners.

CE-2 (reverse item)

Our organization's sales have performed poorly, if compared to the competitors'.

CE-3

The financial strength of our organization is satisfactory for our investors/owners.

CE-4

Our organization has proved capable of adapting to market challenges.

CE-5

Our organization has displayed superior operating efficiency in comparison with our competitors in the last 3 years.

References

Akyuz, G. A., Rehan, M., (2009). Requirements for forming an 'e-supply chain'. International Journal of Production Research, Volume 47, Issue 12, Pages 3265-3287.

Chandra, C., Kumar, S., (2001). Enterprise architectural framework for supply-chain integration. Industrial Management & Data Systems, Volume 101, Issue 6, Pages 290 – 304

Chen, H. and Chen, T. (2003). Governance structures in strategic alliances: transaction cost versus resource-based perspective. Journal of World Business, Spring 38 (1): 1.

Combs J.G., Ketchen D. (1999). Explaining Interfirm Cooperation and Performance. Toward a reconciliation of predictions from the resource-based view and organizational economics. Strategic Management Journal, 20:867-888.

Ford D., Håkansson H., Gadde, LE and Snehota I (2003). Managing Business Relationships, second edition. Chichester: John Wiley & Sons.

Hannan, M.T. and J. Freeman (1989) Organizational Ecology. Cambridge, MA: Harvard University Press

Håkansson H. and Snehota I. (1989). No business is an island. Scandinavian Journal of Management Studies, 4 (3): 187-200.

Håkansson, H., Ford, D., Gadde, L-E., Snehota I, and Waluszewski A.(2009), Business in Networks Chichester: Wiley.

Hillman, A. J., Withers, M. C. and B. J. Collins (2009). Resource dependence theory: A review. Journal of Management 35: 1404-1427.

Jagdev, H. S., & Browne, J. (1998). The extended enterprise-a context for manufacturing. Production Planning & Control, 9(3), 216-229.

Kadushin, C. (2012). Understanding Social Networks: Theories, Concepts, and Findings. Oxford University Press.

Khanwalla P.N., Mehta K. (2004). Design of Corporate Creativity. Vikalpa: The Journal for Decision Makers; Jan-Mar2004, Vol. 29 Issue 1, p13-28.

Lee, H. L., Padmanabhan, V., & Whang, S. (1997). Information distortion in a supply chain: the bullwhip effect. Management science, 43(4), 546-558

McFarland, R.G., Bloodgood, J.M., Payan, J.M., (2008). Supply Chain Contagion. Journal of Marketing, Volume 72, Issue 2, Pages 63-79.

Melnyk, S. A., Rodrigues, A., & Ragatz, G. L. (2009). Using simulation to investigate supply chain disruptions. In Supply Chain Risk (pp. 103-122). Springer US.

Meyer, J. W., & Rowan, B. (1977). Institutionalized organizations: Formal structure as myth and ceremony. American journal of sociology, 340-363.

Naylor, J.B., Naim, M.M., Berry, D., (1999). Leagility: Integrating the lean and agile manufacturing paradigms in the total supply chain. International Journal of Production Economics, Volume 62, Issues 1-2, Pages 107-118.

Ostrom E. and Walker J. (2005). Trust and reciprocity: Interdisciplinary Lessons for Experimental Research. Russel-Sage Foundation.IIASA Interim Report, IR-05-079. IIASA Studies in Adaptive Dynamics, n.111.

Powell, W. W., Koput, K. W., & Smith-Doerr, L. (1996). Interorganizational collaboration and the locus of innovation: Networks of learning in biotechnology. Administrative science quarterly, 116-145.

Ricciardi F., Lombardi P. (2010), Widening the Disciplinary Scope of eParticipation. Reflections after a Research on Tourism and Cultural Heritage. In: Tambouris E., Macintosh A., Glassey O. (eds), Electronic Participation. Second IFIP WG 8.5 International Conference, ePart 2010. Lausanne, Switzerland, August/September 2010. Proceedings. Springer

Sæbø Ø., Rose J., Skiftenes Flak L. (2008): The shape of eParticipation: Characterizing an emerging research area.Government Information Quarterly,Volume 25, Issue 3, July 2008, pp. 400-428

Suchman, M. C. (1995). Managing legitimacy: Strategic and institutional approaches. Academy of management review, 20(3), 571-610.

Sun, H., Yau, H. K., & Ming Suen, E. K. (2010). The Simultaneous Impact of Supplier and Customer Involvement on New Product Performance. Journal of technology management & innovation, 5(4), 70-82.

Turker D. (2009). Measuring Corporate Social Responsibility: A Scale Development Study. Journal of Business Ethics (2009) 85:411–427.

Williamson, O. E. (1985). The Economic Institutions of Capitalism: Firms, Markets, Relational Contracting. New York, NY: Free Press.

Woolley, A. W., Chabris, C. F., Pentland, A., Hashmi, N., & Malone, T. W. (2010). Evidence for a collective intelligence factor in the performance of human groups. Science, 330(6004), 686-688.

Zingales, L. (2009). Capitalism after the crisis. National Affairs, 1, 22-35.

3. Sustainable Altruism for Business Networking: The Costs of Fairness and Cooperation

Abstract   This chapter seeks to contribute to a better understanding of the role of altruism in business relationships. A framework on altruism-enabled behaviors is built and presented, based on game theories, systems approach, and evolutionary studies, and it is discussed in the light of some key management theories (namely, the theory of the population ecology of organizations, the resource dependence theory, the resource based - relational view of the firm, the collaborative network theory, and the institutional theory). Several forms or types of altruism are identified; further, correlated concepts, such as Cooperation, Sharing, Help, Respect, Punishment, Fairness, Pride, Gratitude, and Reputation, are defined and discussed. The concept of "sustainable altruism" is introduced and discussed. Finally, a Construct and related Scale is developed, "Business Network Strength", aimed to measure the extent to which an organization's business network effectively leverages altruism-enabled mechanisms to make long-term inter-organizational fairness and cooperation sustainable.

3.1 Introduction

There is growing interest towards economic interactions occurring without, or beyond, explicit contracts, regulatory institutions or rational market negotiations (Sigmund, 2002). Most of these basic economic interactions are triggered by emotions and feelings, such as gratitude, pride, sympathy or indignation, that are usually overlooked in business settings analysis (Nowak et al, 2000); nevertheless, behaviors occurring during these "primitive" interactions often result in surprising self-organizing effects (Fehr et al., 2002).

As a consequence, many scholars suspect that such down-to-earth interactions may have a key role in real-world business settings, even if classical economic theory disregards them as marginal (Simon, 1984). This chapter is aimed to define the state-of-the art in academic understanding of *altruism*: a phenomenon which may be considered incomprehensible if one starts from the classical rational actor assumption (Nowak et al., 2000).

This research goal was pursued through an inter-disciplinary approach, investigating literature focusing on altruism in the following fields: (a) game theories, systems theories and evolutionary studies; and (b) management theories.

This literature analysis was aimed to find out how the research communities focusing on the fields above, (a) and (b), answered the following *Research Questions:*

- *What is altruism and why does it exist? (Paragraph 3.2)*
- *What is the role of altruism in inter-organizational business interactions? (Paragraph 3.3)*
- *Under what conditions are altruistic behaviors sustainable? (Paragraph 3.4)*

3.2 What Is Altruism and Why Does It Exist?

3.2.1 Altruism-Related Concepts

An extremely simple definition of altruism is adopted here, borrowed from the evolutionary approach (described in the Literature Analysis below): *altruism occurs when an actor chooses to pay a cost for someone else to gain a benefit.* This definition entails no moral implication; more specifically, the motivations for benefitting others are not considered relevant to distinguish altruistic from non-altruistic behaviors.

The costs of altruism can consist in:

1. *lost opportunities* (the altruist actor renounces an opportunity to increase ownership and/or exploitation and/or control of valuable resources, i.e. to maximize payoff and/or to minimize losses);
2. *lost resources* (the altruist actor gives part of his/her/its own resources away, renouncing resource exploitation and/or ownership, but not necessarily renouncing control on how these resources will be used by others);
3. *lost control* (the altruist actor shares part of his/her/its own goals and resources with other actors, renouncing total control on goal setting, on the one side, and on how and why shared resources will be used, on the other side; but not necessarily renouncing exploitation and/or ownership of the shared resources).

In cases in which the cost of altruism is an opportunity loss, altruism can be considered as opposed to *opportunism*; in fact, opportunism occurs when an actor chooses to maximize his/her/its own immediate payoff (and/or minimize losses), even by breaking rules, taking advantage of other actors' weakness and/or abusing their trust.

In cases in which the cost of altruism is a resource loss, altruism can be considered as opposed to *selfishness*; in fact, selfishness occurs when an actor refuses to give owned resources away for someone else to gain a benefit.

In cases in which the cost of altruism is a control loss, altruism can be considered as opposed *independency*; in fact, independency occurs when an actor refuses to give up control on decisions and/or resources for someone else to gain a benefit.

Three key altruism-enabled behaviors were identified through the literature review that will be described in the paragraphs below. These three altruism-enabled behaviors are defined consistently with the framework on the costs of altruism, as the reader can see:

1. Altruistic *Respect* implies that the actor abides by agreements and rules the other actors rely on, even if an opportunistic violation of such agreements or rules would result in higher immediate payoff. In other words, the concept of Respect will be used here to identify a behavior where opportunism, although possible, is given up.
2. Altruistic *Help* implies that the actor dedicates part of his/her/its own resources to pursue (also) another actor's goals. In other words, the concept of Help will be used here to identify a behavior where selfishness, although possible, is given up.

3. Altruistic *Sharing* occurs when the actor is ready to renounce total autonomy in decision-making and total control on a certain resource so that other actors are allowed, to some extent, to decide whether, how and why use the shared resource. In other words, the concept of Sharing will be used here to identify a behavior where independency, although possible, is given up.

The behaviors of Respect, Help and Sharing partially overlap. In fact, there are behaviors that imply renouncing both opportunities and resource ownership (Respect + Help); there are behaviors that imply renouncing both opportunities and control on goals and resources (Respect + Sharing); there are behaviors that imply renouncing both resource ownership and decision / resource control (Help + Sharing) and there are behaviors that imply giving up opportunities, resource ownership and control (Respect + Help + Sharing).

On the other hand, there are cases in which the cost that the actor is ready to pay consists in lost opportunities only, or in lost resource ownership only, or in lost control only.

Figure 3.1 illustrates this articulated "map" of the possible types and costs of altruism.

In case the actor renounces opportunism, but without helping and sharing, the receiver is able to rely at least on the respect of rules and agreements, which is extremely important in business settings: this area of altruistic behavior is defined here as *Fairness*.

The area of altruistic behavior where Respect is associated to Help and/or Sharing is defined here as *Cooperation*.

There may also be behaviors of Opportunistic Help and/or Opportunistic Sharing, in case the actor does not renounce opportunism, even while helping others and/or sharing resources and decisions with them. This area is defined here as altruistic *Manipulation*.

Figure 3.1 illustrates all these possibilities.

As the reader can see, even if the definition of Altruism proposed here is quite simple, it describes a complex and multi-faceted phenomenon.

The model proposed here classifies altruistic behavior as for the cost the actor is ready to pay (Respect, if the cost is renouncing opportunities; Help, if the cost is renouncing resources; Sharing, if the cost is renouncing Control; see Figure 3.1) and as for the relational environment it creates (Fairness, if the behavior consists in Respect only; Cooperation, if the behavior consists in both Respect and Help/Sharing; Manipulation, if the behavior consists in Help/Sharing without Respect).

But why does this multi-faceted phenomenon exist, and more specifically why does it exist in business settings?

If we concentrate on a single interaction, altruism does not make sense. But researches focusing on long-term interactions provided us with many, and of course multi-faceted, interesting explanations.

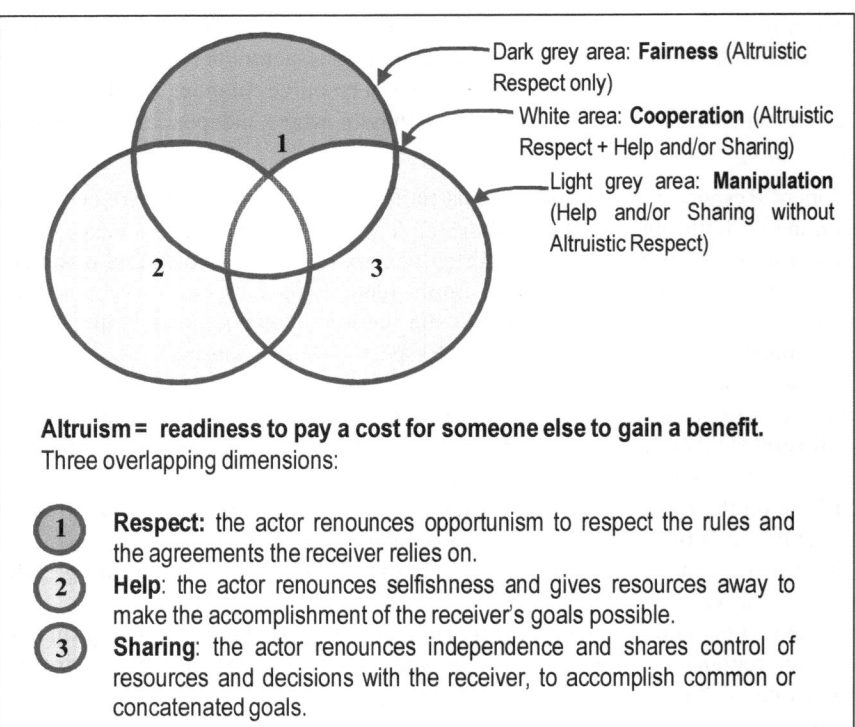

Altruism = readiness to pay a cost for someone else to gain a benefit.
Three overlapping dimensions:

1. **Respect:** the actor renounces opportunism to respect the rules and the agreements the receiver relies on.
2. **Help:** the actor renounces selfishness and gives resources away to make the accomplishment of the receiver's goals possible.
3. **Sharing:** the actor renounces independence and shares control of resources and decisions with the receiver, to accomplish common or concatenated goals.

Fig. 3.1 Altruism and Related Concepts

3.2.2 Altruism in Game Theories

Game theories are aimed to understand economic interactions between humans. This approach starts from designing a game, i.e. a " highly abstract, and sometimes even contrived instance (...) of interactions between independent decision makers" (Sigmund, 2002). Then, the researcher tries to predict, on the basis of given assumptions, how humans will play the game. Finally, the predictions are tested in experimental settings, and as a consequence the assumptions are either corroborated, or proven false (Bolton et al., 1998; Andreoni and Miller, 1993).

One of the most famous examples is the Ultimatum Game. "In the Ultimatum Game, two players are offered a chance to win a certain sum of money. All they must do is divide it. The [player chosen as] proposer suggests how to split the sum. The responder can accept or reject the deal. If the deal is rejected, neither player gets anything" (Nowak et al., 2000). The two players. i.e. the Proposer and the Responder, cannot negotiate nor communicate before the offer.

Classical game theory firstly addressed the Ultimatum Game on the basis of the classical economic assumption of rational opportunism of players. It then predicted that the proposer would offer the lowest possible sum, for example one dollar out of 100, and that the respondent would accept, because from a "rational" standpoint even one dollar is better than zero. But this is not what happens in the real world. The Ultimatum Game has been replicated hundreds of times, in different experimental conditions, involving people of all social conditions, all cultures and all geographic origins, and the outcome is always robustly the same (Nowak et al. 2000): most proposers offer a sum which is much higher than the lowest possible, between 25% and 45% of the total amount; and most responders who are offered a sum lower than 20-30% reject.

In other words, this experimental game, which was created under the assumption of rational opportunism, paradoxically showed that people tend to behave altruistically, to expect the others to do so, and to indignantly punish opportunists, even at the cost of losing payoff.

As a consequence of this failure in predictions, there was a sort of copernican revolution in game theories. Scholars started thinking in terms of populations instead of individuals; in terms of embedded emotional triggers instead of rational opportunism; in terms of long-term trial-error strategies instead of short-term aware planning. In other words, the evolutionary thought arrived on the game theory scene.

The first model based on this novel approach was proposed by Nowak and Sigmund (1992, 1993). Since then, a great amount of theoretical and experimental work has been proposed in this field, which is one of the most viable in economic studies, as we will see below.

According to this approach, an interaction strategy (e.g. a certain reciprocation rule, or the punishment of opportunists) spreads in a given population if it is successful; but "success" cannot be measured on the basis of the single interaction. Complex mathematical models were developed to predict what behaviors are successful in the long run, at given conditions. These mathematical models, differently from those based on the rational actor assumption, resulted in successful predictions of actual altruistic behaviors of players in games such as the Ultimatum Game and many others (Nowak, 2006). To develop these mathematical models, evolutionary game theory rooted in system theory, to which the next paragraph will be dedicated.

What is, then, altruism, from the evolutionary game theory standpoint? All the writings imply a consistent idea of altruism, that is synthesized as follows. *Altruism is a strategy for economic interaction that may result, at certain conditions, in a more stable and more successful equilibrium in a given population, than that resulting from an undisputed invasion of egoism* .

3.2.3 *Altruism in Systems Theories*

System thinking is a vast phenomenon which has been deeply influencing studies in almost all fields in the last decades. We will limit ourselves to briefly presenting those branches of system theories that seem more interesting to understand the phenomenon of altruism in business settings: (i) complex adaptive systems, also called adaptive dynamics, and (ii) social rule systems, also called institutional systems.

Adaptive dynamics studies focus on how systems evolve as a consequence of different interaction dynamics within them; for example, a system may collapse for an epidemic of opportunistic behavior, which prevents actors from cooperating on a common problem (Kryazhimskiy and Kleimenov, 1998).

Whilst Adaptive Dynamics Studies rest on the assumption that the system is self-organizing and evolves for a series of spontaneous trial-error activities during interactions (Young, 1996), Institutional Systems studies focus on how interactions may be, to some extent, routed or controlled by a governance system. Institutional systems, for example, are developed for the governance of common pool resources, such as security or fisheries, in order to enforce mechanisms preventing over-exploitation of resources or dangerous defection rates in cooperation (Ostrom and Walker, 2005).

In all system theories, the unit of analysis is not the individual part of the system, but the system of long-term *interactions* between parts of the system. Game modeling is then a natural complement of system thinking. In this field of studies, altruism is a key concept: "We will show that the interaction between selfish and strongly reciprocal individuals is decisive for the understanding of human cooperation. We identify conditions under which selfish individuals trigger the breakdown of cooperation, and conditions under which the strongly reciprocal individuals have the power to ensure wide-spread cooperation" (Fehr and Fischbacher, 2003).

Consistently with the game theory approach, Adaptive Dynamics and Institutional Systems Studies see altruism as a strategy competing with opportunism in each individual interaction. System theories analyze altruistic behavior without any moral assumption (West et al., 2007). Altruism may win, but not because it is better or nicer or fairer: it wins when it ends up yielding a higher payoff. These studies demonstrate that a certain degree of opportunism tends to be successful in the short term, whilst a certain degree of altruism tends to be successful in the long run. But what is this "certain degree" of altruism? System theories allow us to build sound mathematical models and experimental evidence to answer this question in different settings (Hofbauer and Sigmund, 2003). For example, it has been demonstrated that the possibility of effectively punishing defectors is often an essential condition for cooperative behaviors to enduringly spread within a population (Sigmund, 2007; Hauert et al., 2007). Consistently, people are instinctively willing to punish opportunists, even if punishment is costly; this kind of costly punishing is called "altruistic punishment" because it is proven beneficial for the collectivity, in that it prevents the invasion of cheaters.

In a nutshell, the main contribution of system theories to understanding the phenomenon of altruism is a precise, mathematical identification of some important conditions under which the altruistic behavior is more successful than the opportunistic behavior. These "conditions for altruism" will be more thoroughly described in Paragraph 3.3; system thinking provides important insights into the relationships between altruism and phenomena like punishment, reputation, negotiation, spatial proximity, gossip, competition, or even war.

3.2.4 Altruism in Evolutionary Studies

Evolutionary studies provide powerful frameworks to understand the phenomenon of altruism (Axelrod and Dion, 1988). This could be surprising, at a first glance; in fact, classical Darwinian theory is based on competition and selfishness. But the evolutionary thought soon realized that competition does not take place at one level only (West et al., 2007). For example, if interaction at a certain level is competitive and selfish (e.g. a war between two anthills) a cooperative interaction will be needed at lower levels (e.g. mutual help among ants of the same anthill). "The two fundamental principles of evolution are mutation and natural selection. But evolution is constructive because of cooperation. New levels of organization evolve when the competing units on the lower level begin to cooperate. Cooperation allows specialization and thereby promotes biological diversity. Cooperation is the secret behind the open-endedness of the evolutionary process. *Perhaps the most remarkable aspect of evolution is its ability to generate cooperation in a competitive world. Thus, we might add "natural cooperation" as a third fundamental principle of evolution beside mutation and natural selection*" (Nowak, 2006; my emphasis).

Cooperation is then a key strategy in nature, and when cooperation requires sacrifice, then altruistic attitudes are selected as an alternative and antidote to egoistic attitudes, to make cooperation possible (Sigmund, 1998).

In evolutionary studies, cooperation is studied in that it is costly, and then the definitions of altruism and cooperation tend to overlap: according to evolutionary view, (altruistic) cooperation occurs when an individual pays a cost, c, for someone else to gain a benefit, b. Costs and benefits are measured in terms of fitness (Nowak, 2006). The values of b and c are included in mathematical models, which seek to predict social behaviors.

In some cases, altruistic behaviors are almost automatic: for example, ants don't hesitate about risking their lives to protect their anthill; many mammals don't hesitate about risking their lives to protect their offspring. In other more complex cases, such as the economic games analyzed above, an evenly match between altruistic and selfish strategies takes place: both attitudes may reveal successful, according to circumstances, and so they are both basic to the brain and hard-wired in feelings and emotions. "We feel fine if we help others and share with them. But where does this 'inner glow' come from? It has a biological function. We eat and love because we enjoy it; but behind the pleasure stands the evolutionary program commanding us to survive and to procreate. In a similar way, social emotions like friendship, shame, generosity or guilt act to prod us towards achieving biological success in complex social networks" (Sigmund 2002).

Altruistic cooperation, in sum, involved the co-evolution of a large range of emotions (which we share with many social animals): e.g. hypocrisy, suspicion, friendship, sympathy, guilt, sense of honor, sense of belonging, etc. Such emotional triggers have great importance in business settings, much beyond the rational actor assumption.

Evolutionary studies witnessed a great dynamism with regards to theories of altruism in the last 15 years. In the 1990s, it was commonly accepted that genuine evolutionary processes could explain only certain forms of altruism, and namely: kin altruism (I help only my relatives, because even if I die they will carry my genes on); reciprocal altruism (I help only who is likely to help me in turn); and some particular forms of altruism aimed at being chosen as a mate in sexual competition (I help the best potential mates and so they will probably choose me) (Roberts, 1998). But in the last years, the growing powerful integration between evolutionary theories and game/system approaches led to understand that also more complex forms of altruism are well explained by evolutionary processes.

3.2.5 Different Forms of Altruism in Integrated Evolutionary-System-Game Theories

The strong cross-fertilization between evolutionary studies, game theory and system theory led to an impressive widening of investigation focus as for the phenomenon of altruism in the last 15 years. Even if there is still some resistance (see for example West et al. 2007), nowadays evolutionary theories can provide sound explanations also to altruistic cooperation between non related and non-reciprocating individuals, and can predict acts of generosity that are not directed to reproductive success.

So far, the integrated evolutionary-systems-game theory approaches identified several typical receivers of altruistic behaviors, that correspond to different specific reasons to behave altruistically. These different reasons result in different forms of altruism, which are summarized below:

1. **Kin Selection**: I respect / help / share decisions and resources with my kin group, and I punish my kin's behaviors if unfair to the family. This enhances the probabilities that my genes will propagate (West et al., 2007; Sigmund, 2007).
2. **Direct Reciprocity**: I respect / help / share decisions and resources with those who are likely to be in the condition to interact with me in the near future. Moreover, I punish them, if they behave unfairly. This enhances the probabilities that I will be respected / helped / allowed to access shared resources or decisions when I need it (Trivers, 1971).
3. **Indirect Reciprocity**: I respect / help / share decisions and resources with others *even if* they are *not* likely to interact with me in the future, for the sole purpose to build my name as a good cooperator (Nowak and Sigmund, 2005). Moreover, I punish them, if they behave unfairly, to let it be known that it is not convenient to behave opportunistically against me (Sigmund, 2007). This enhances the probabilities that I will be respected / helped/ allowed to access shared resources or decisions in turn by those who know my good *reputation* (Brandt et al., 2006). "Often the interactions among humans are asymmetric and fleeting. One person is in a position to help another, but there is no possibility for a direct reciprocation. We help strangers who are in need. We donate to charities that do not donate to us. *Direct reciprocity is like a barter economy based on the immediate exchange of goods, whereas indirect reciprocity resembles the invention of money. The money that fuels the engines of indirect reciprocity is reputation.* Helping someone establishes a good reputation, which will be rewarded by others. [...] Natural selection favors strategies that base the decision to help on the reputation of the recipient. *Theoretical and empirical studies of indirect reciprocity show that people who are more helpful are more likely to receive help.* " (Nowak, 2006, my emphases).

4. **Group Selection**: if the group I belong to is subject to competition with other groups, I respect / help / share decisions and resources with my group's members and I reliably respect my group's rules. This will enhance my group's fitness (Rainey and Rainey, 2003). Moreover, I punish unfair behaviors within my group, even if it is costly to me, this will make epidemics of unfair behaviors in the population I belong to less likely; as a consequence, the entire population will benefit from my behavior (Sigmund, 2007).
5. **Resource-Oriented Altruism:** I respect / help / share decisions and resources with actors that can allow me access to valuable resources, and/or whose cooperation can protect or increase the value and/or successful exploitation of the resources I already own and/or use. (Ostrom and Walker, 2005).
6. **Goal-Sharing Altruism:** I respect / help / share decisions and resources with actors whose interests and/or goals are common or interdependent with mine. By doing so, I make the successful achievement of my goals more likely; for example, if I help my debtor to find a better job, it is more likely that my credit will be regularly paid back, even if I keep my help secret and no gratitude or reciprocation process is then involved (Gintis, 2000).
7. **Competitive Altruism:** I respect / help / share decisions and resources with actors that can decide about my competitive position, for example my boss or my customers (but I often dare not to punish them even if they are unfair, because of their position of power). Generous behaviors demonstrate that I can afford the costs of altruism and that I could be a valuable cooperator; then, generous behaviors may enhance the probabilities that I will be chosen e.g. as a mate, as a partner, as an employee, as a supplier, as a leader, overcoming my competitors (Gintis et al., 2001; Roberts, 1998).
8. **Unconditional Altruism:** I respect/help/ share decisions and resources with someone else, for none of the reasons mentioned above: for example, I secretly help a stranger I will never see any more, so that I gain no payoff, not even reputation. This happens because the strong emotional triggers evolved to make "useful" altruism possible may reward me also for "useless" altruistic behaviors, making fair behaviors pleasurable, per se, especially if feelings of sympathy, love or affection are involved (Sigmund, 2002).

Of course, these eight different reasons to behave altruistically may overlap and mix in real world situations. For example, in business networks we can find both forms of competitive altruism and of indirect reciprocity. In family businesses, both kin selection and direct reciprocity may come into action. In case of war, the dramatic external threat may enhance the effect of altruistic punishment to fuel group selection, and so on.

3.3 What Is the Role of Altruism in Inter-Organizational Business Interactions?

The term "altruism" is almost absent in studies focusing on the nature and organization of firms. A search in the Google Scholar database in June, 2012, using "organizational AND altruism" and then "firm AND altruism" as keywords, yielded no relevant writing among the first 50+50 results: it seems that the world "altruism" is included in management studies only in case of specific issues, such as charity organizations or family businesses. In other words, altruism is generally not considered as a metric to assess an organization's value, nor as a basic concept to understand what a firm is, and why does it survive.

On the other side, there is a growing interest in altruism-related concepts, such as *cooperation, fairness, reliability/accountability/respect* (see Paragraph 3.2.1). Different organizational theories tend to focus on different aspects of altruism and cooperation: a synthetic overview of 5 main theoretical approaches to business networks follows.

3.3.1 Altruism in the Theory of the Population Ecology of Organizations

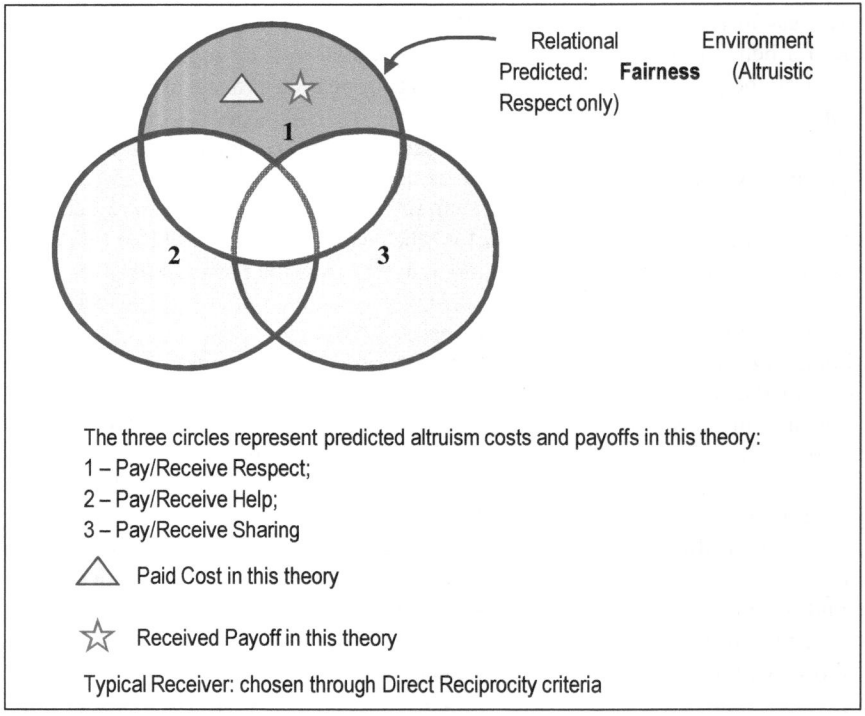

The three circles represent predicted altruism costs and payoffs in this theory:
1 – Pay/Receive Respect;
2 – Pay/Receive Help;
3 – Pay/Receive Sharing

⚠ Paid Cost in this theory

☆ Received Payoff in this theory

Typical Receiver: chosen through Direct Reciprocity criteria

Fig. 3.2 Inter-Organizational Altruism According to the Theory of the Population Ecology of Organizations

In the theory of the Population Ecology of Organizations, inter-organizational networks are seen as populations of competing firms where change and innovation are mainly due to the entry and selective replacement of organizations. Organizational founding, growth, change and mortality are at the core of this theoretical approach's interests (Carroll and Hannan, 2000). Inter-organizational cooperation is not taken into consideration as a possible predictor of improved organizational fitness or innovation capabilities. This is not surprising, since this approach dates back to the 1970s (Hannan and Freeman, 1977), and is then based on classical evolutionary thought, which in effect, before the 1990s, was not aware of the essential evolutionary role of cooperation (see above, Paragraph 3.2.4). On the other hand, this theory includes claims on Organizational Inertia and Organizational Change, that are interestingly consistent with the theory of direct and indirect reciprocity (Paragraph 3.2.5). In fact, the Population Ecology Theory claims that organizations that are reliable and accountable are favored by selection in the business environment, and are then more likely to survive. On the other hand, successful reliability and accountability have a negative side effect: they result in a high degree of organizational inertia and resistance to change.

Reliability and accountability can be seen as aspects of *Altruistic Respect* (see above, Paragraph 3.2.1). This theory, in other words, explains altruistic behaviors of inter-organizational Respect in non-cooperative business settings. The Population Ecology of Organizations predicts that well-established organizational reliability and accountability in non-cooperative business networks enhances fitness as long as change is not needed; but when innovation becomes necessary, the process of change itself is so disruptive in these highly-inertial settings, that it is likely to result in organizational mortality (Ulrich, 1987; Hannan et al., 2007). In this theory, then, only behaviors of altruistic Respect are taken into consideration; Direct Reciprocity is the only criterion for choosing the receivers; and becoming the receiver of reciprocated Respect is the expected long-term payoff of this type of altruism. The relational environment predicted by this theory can be described as Fair within the framework adopted in Paragraph 3.2.1 (Figure 3.2).

3.3.2 Altruism in the Resource Dependence Theory

In the Resource Dependence Theory, inter-organizational networks are seen as means for controlling extra-organizational critical resources. According to this theoretical approach, an organization's main goal when interacting with other organizations is to reduce uncertainty about third-party owned resources (Pfeffer and Salancik, 1978). As a consequence, this approach concentrates on formalized agreements, such as long-term outsourcing contracts, inter-firm vertical integration, supply chain integrated management, joint-ventures; and on non-formalized strategies, such as lobbying, board interlocks, inter-organizational bullying, market power games. In other words, this approach sees cooperation as possible only if either a formal, sound inter-organizational agreement is present, like in outsourcing contracts, or the common goal is to exercise power, like in lobbying activities (Hillman et al., 2009). Selfishness and opportunism play an important explanatory role in this theory, which is highly compatible with the Transaction Cost Theory (Williamson, 2009), and is not interested in the possible beneficial effects of long-term informal cooperation. The only altruism-enabled behavior involved by this theory is the Respect of rules and agreements (see above, Paragraph 3.2.1), like in the Population Ecology Theory; but, differently from the Population Ecology Theory, the Resource Dependence Theory does not predict counter-productive effects of this behavior in the long run.

In this theory, then, only behaviors of altruistic Respect or Help are taken into consideration. Resource Oriented Altruism, Goal Sharing Altruism and Competitive Altruism (see above, Paragraph 3.2.5) are the criteria for choosing the receivers, according to this theory; and becoming the receiver of resource Sharing is the expected long-term payoff of this type of altruism. There is an asymmetry between the cost that the actor is ready to pay and the benefit that expects (see above, Paragraph 3.2.1): consistently, the intra-network relational environment predicted by this theory can be described as Fair / Cooperative only in case of lobbying or strongly formalized alliances, whilst it tends to be Manipulative with suppliers, customers, and value chain partners (Figure 3.3).

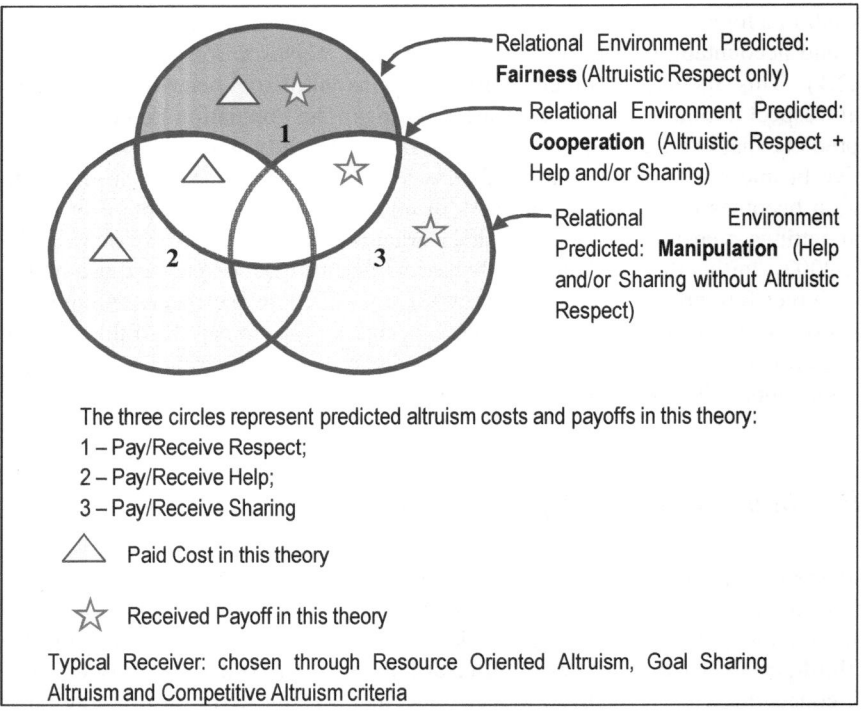

Fig. 3.3 Inter-Organizational Altruism According to the Resource Dependence Theory

3.3.3 Altruism in the Resource Based / Relational View of the Firm

In the *Resource Based View of the Firm*, inter-organizational networks are seen as means to exploit the heterogeneity and complementarity of firms, so that each actor can access strategic resources beyond its own boundaries, in order to gain competitive advantage (Combs and Ketchen 1999). This approach explains cooperation and altruism only between heterogeneous organizations, which can exchange valuable, rare, inimitable and/or non-substitutable resources (Chen and Chen 2003). In this perspective, cooperation occurs as a strategy to accumulate resources: firms should select partners that prove capable to enrich and/or complement their resource base (Fink and Kessler 2010). An extension of the Resource Based View, the *Relational View of the Firm*, goes beyond that and suggests to consider the cooperative network, per se, as a source of long-term competitive advantage (Lavie, 2006). According to this view, there can be relation-specific assets and knowledge sharing routines that, if effectively governed, result in an idiosyncratic system of inter-firm linkages, impossible to imitate by competitors. If that is the case, it is the relationship with a key supplier, for example, that ensures competitive advantage, much more than the product purchased from that supplier.

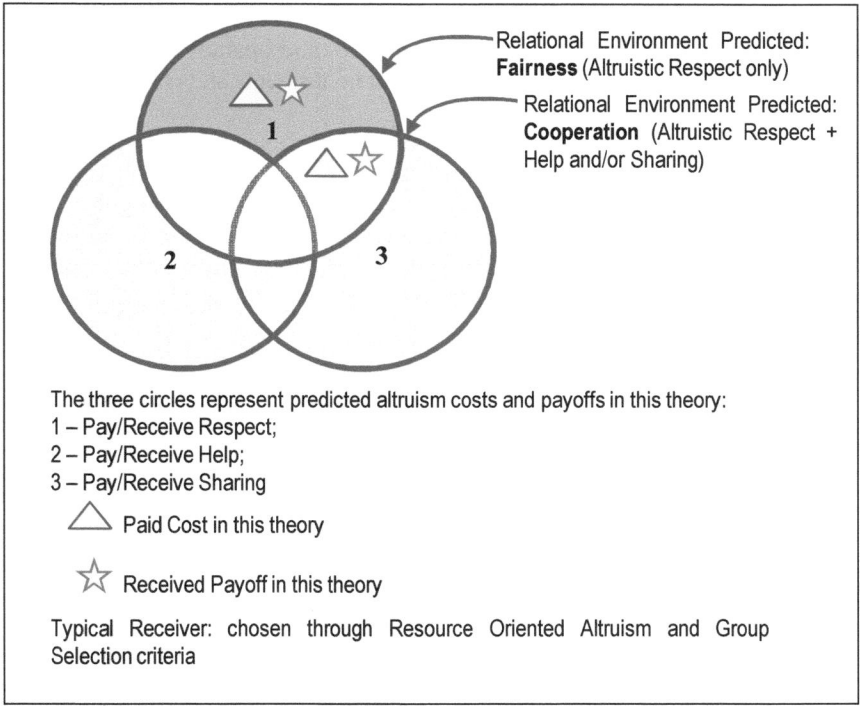

The three circles represent predicted altruism costs and payoffs in this theory:
1 – Pay/Receive Respect;
2 – Pay/Receive Help;
3 – Pay/Receive Sharing

△ Paid Cost in this theory

☆ Received Payoff in this theory

Typical Receiver: chosen through Resource Oriented Altruism and Group Selection criteria

Fig. 3.4 Inter-Organizational Altruism According to the Resource Based / Relational View of the Firm

In this theory, then, only behaviors of altruistic Respect or Sharing are taken into consideration. Resource Oriented Altruism and Group Selection (see above, Paragraph 3.2.5) are the criteria for choosing the receivers, according to this theory; and becoming, in turn, the receiver of resource Sharing is the expected long-term payoff of this type of altruism. The relational environment predicted by this theory can be described as Fair / Cooperative (see above, Paragraph 3.2.1) (Figure 3.4).

3.3.4 Altruism in the Collaborative Network Theory

In the *Collaborative Network Theory*, inter-organizational networks are seen as means to share goals, interests, resources, risks, costs and opportunities (Greenhalgh, 2001). According to this approach, the attempts to control external resources through power games, lobbying, abuse of position of strength, or also through detailed long-term contracts and agreements, are counter-productive in the long run. Successful organizations, on the contrary, establish flexible strategic relationships, based on cooperation (shared goals and resources), loyalty, fairness, trust and reciprocal engagement. These informal alliances allow flexible co-adaptation and continuous fine-tuning of goals, strategies and operative interactions, on the one side, while on the other side encourage long-term investments and co-innovation efforts.

In this theory, then, all behaviors of altruistic Cooperation are taken into consideration, i.e. both Help and Sharing. Direct and Indirect Reciprocity, Resource Oriented Altruism and Goal Sharing Altruism (see above, Paragraph 3.2.5) are the criteria for choosing the receivers, according to this theory; and becoming, in turn, the receiver of cooperative behavior is the expected long-term payoff of this type of altruism. The relational environment predicted by this theory can be described as highly Cooperative (see above, Paragraph 3.2.1) (Figure 3.5).

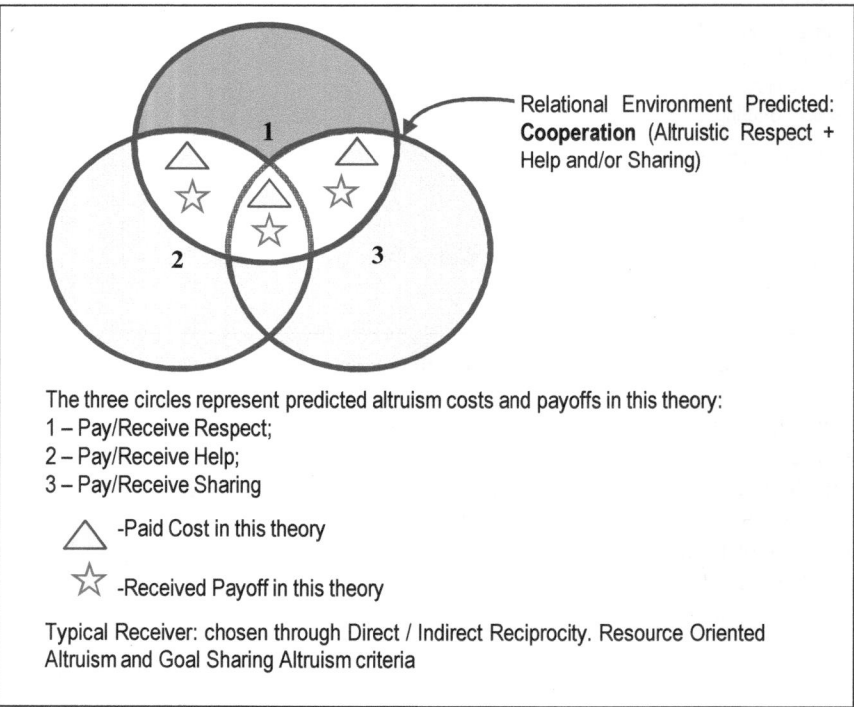

Fig. 3.5 Inter-Organizational Altruism According to the Collaborative Network Theory

3.3.5 Altruism in the Institutional Theory

The *Institutional Theory* sees business networks as means to gain legitimacy in a certain social context and/or business community.

According to this approach, organizations seek to comply with the social rules and expectations that can give them prestige and reputation. They behave altruistically to demonstrate that they deserve legitimacy; behaviors conform to expectations through mimetic, coercive and normative pressures.

In this theory, then, all behaviors of altruistic Cooperation are taken into consideration, i.e. both Help and Sharing, along with, of course, Fairness. Indirect Reciprocity and Competitive Altruism (see above, Paragraph 3.2.5) are the criteria for choosing the receivers, according to this theory; whilst the long-term expected payoff for altruistic behaviors is outside the circles of Altruism, and consists in Legitimacy. The relational environment predicted by this theory can be described as Fair and Cooperative (see above, Paragraph 3.2.1) (Figure 3.6).

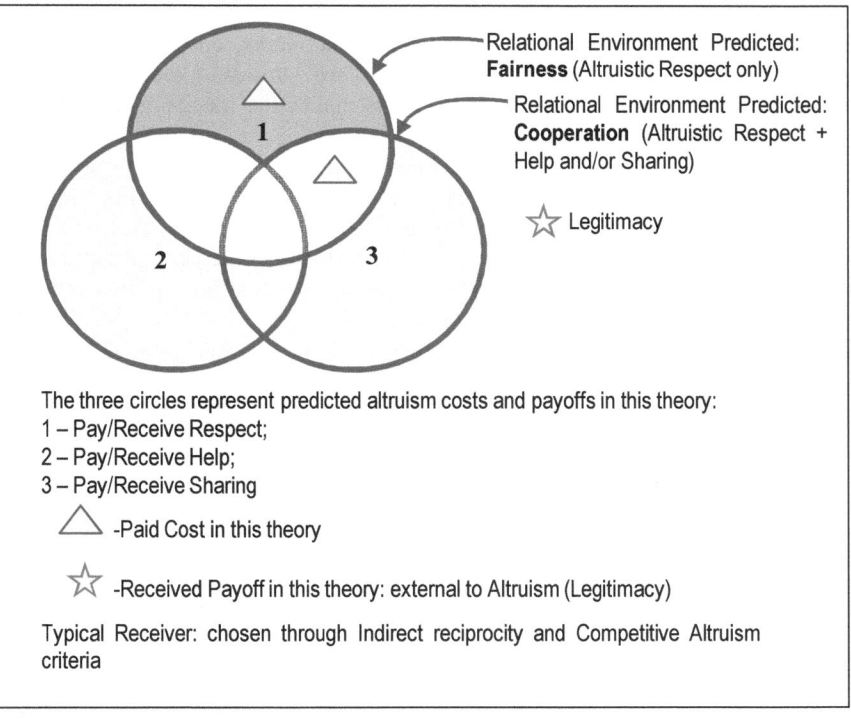

Fig. 3.6 Inter-Organizational Altruism According to the Institutional Theory

3.4 Under What Conditions Are Altruistic Behaviors Sustainable?

Fairness and cooperation are continuously cited as key enablers for network performances.

From an evolutionary point of view, altruism is an interesting strategy, since whereas it may reduce the single individual's fitness, it is likely to dramatically enhance the overall fitness of the cooperating group. But this does not mean that altruistic fairness and cooperation are always and in any case positive phenomena. Evolutionary and systems theories are developing an impressive amount of studies identifying the conditions under which altruistic cooperation is more sustainable, and then more likely to be successful (Ricciardi and Cantù, 2011). On the basis of such literature, it can be stated that, in business network contexts, at least one of the following four conditions must come about to make long-term inter-firm fairness and/or cooperation emerge:

1. The first condition is the *non-extemporary nature of interactions*. If business actors *often interact with the same partners*, this makes the mechanism of *direct reciprocity*, driven by feelings like honor and gratitude, come into action (see above, Paragraph 3.2.5).
2. The second condition is *transparency*. In transparent networks, cooperative and non-cooperative behaviors do not remain unknown, but *are witnessed and made known* throughout the population (Brandt et al., 2006). This makes the mechanism of *indirect reciprocity* possible(see above, Paragraph 3.2.5).
3. The third condition is the *sustainability of network size*. If my altruistic behaviors benefit neither a single individual, nor the entire population, but all and only my "network neighbors", this may contribute to the forming of a successful *network cluster* (Lieberman et al., 2005). *Network reciprocity* is more likely to be successful if the average number of *network neighbors* is small, and/or if the additional cost for a further neighbor to receive network benefit is negligible (Ohtsuki et al., 2006).
4. The fourth condition is the presence of *regulatory institutions* working at the network level, capable to detect and to punish opportunistic behaviors, in order to systematically make cooperative behaviors more rewarding, and opportunistic behaviors less rewarding. Regulatory institutions are particularly necessary when cooperation involves the exploitation of *common pool resources*, i.e. common resources that are barely sufficient for everyone's appetite and subject to irreversible drain if abused. I refer here to the so-called "tragedy of the commons", where common resources are greedily over-exploited by actors and end up being destroyed (Ostrom and Walker, 2005).

3.5 Measuring Business Network Strength

In this Chapter, I sought to demonstrate that altruism-enabled behaviors are essential for successful business networking. More specifically, I sought to provide a precise definition of two core altruism-enabled business interaction strategies, i.e. Fairness and Cooperation.

On the other side, I will seek to operationalize the four conditions for altruism described in Paragraph 3.4.

Since altruism-enabled behaviors, although important, are under-investigated in management and organizational literature, I decided to develop a scale to assess it, i.e. "Business Network Strength".

Business Network Strength measures the extent to which an organization's business network effectively leverages altruism-enabled mechanisms to make long-term inter-organizational fairness and cooperation sustainable.

Consistently with the theoretical model developed in Paragraph 3.4, this Construct has been designed with four Dimensions: (1) Interaction Repetitiveness, (2) Network Transparency, (3) Capability to Select and Manage Interactions, and (4) Capability to Manage Conflicts and to Punish Opportunist Behaviors.

The Scale was then discussed and fine-tuned with the five experts involved in this research (see Chapter 1). The results follow.

3.5.1 Business Network Strength (NS)

3+2+7+7 Items Scale

Do you agree with the following statements? Please consider your organization's situation **in the last three years**. *(1=strongly disagree, 5=strongly agree, 3=neutral).*

First Dimension: Interaction Repetitiveness - NS-IR
NS-IR-1
Many business relationships within our Business Network are so trustful and cooperative that it would be hard to replace them, both for us and for our Network Partners.
NS-IR-2
In many relationships within our Business Network, the reciprocal knowledge between Network Partners is so specific and accurate, that both partners can simplify the reciprocal interaction activities, saving money and energies.
NS-IR-3
Within our Business Network, important and effective investments have been made in some relationships, and as a consequence the interruption of such relationships is unlikely.
NS-IR-4
Within our Business Network, there are organizations with divergent interests, that in my opinion are likely to quit.

Second Dimension: Network Transparency - NS-TR
NS-TR-1
During the interactions with our Business Partners, we spend time also telling and commenting on experiences and events occurred to other Business Partners.
NS-TR-2
In our Business Network, the news about a wrong or mistake made by anyone in our milieu immediately spread and are made known to all the Network Partners.

Third Dimension: Capability to Select and Manage Interactions – NS-SM
NS-SM-1
Our organization promotes cooperative and intense communication with those external subjects (e.g. faithful customers, strategic suppliers, banks, associations, etc.) with which recurrent business interactions take place.
NS-SM-2 (reverse item)
In our organization, it is not clear who is expected to contribute to select, and authorize investments on, the most important long-term business relationships.
NS-SM-3
The external subjects with which our organization carries on long-term relevant business relationships are chosen carefully, with the contribution of all the areas involved in each specific relationship (e.g. Sales people, Operations people, Information Systems people).
NS-SM-4
Our organization developed special procedures (e.g. better payment conditions, simplified bureaucracy, top managers' direct commitment) to manage the business relationships identified as strategic or long term ones.
NS-SM-5
Our organization developed specific procedures or criteria aimed at tuning up the risk levels accepted in a certain business relationship according to its trustworthiness and importance.
NS-SM-6
Our organization developed appropriate and effective technological solutions (e.g. information systems, logistic systems) dedicated to optimize long-term business interactions.
NS-SM-7
Our organization has proved capable of successfully designing and managing complex contracts for long-term business relationships, such as outsourcing contracts, joint-ventures, service level agreements, etc.

Fourth Dimension: Capability to Manage Conflicts and to Punish Opportunist Behaviors
NS-CP-1
Within our Business Network, the importance and usefulness of the cooperative relationships between Network Partners are strongly perceived.
NS-CP-2
Within our Business Network, there is a reciprocally positive attitude towards the values and culture of the other Network Partners.
NS-CP-3 (reverse item)
Our Business Network may work better, if only the network Partners had not so different languages and communication styles.
NS-CP-4
Should a Business Partner behave unfairly with our organization, we would not take that lying down even if this carries risks and costs.
NS-CP-5
In our Business Network, who does not behave is cut out from the whole system, or in any case is blamed and punished by all the Network Partners.

NS-CP-6

In our Business Network, some relationships are ruled by special long-term contracts or agreements, such as alliances, joint-ventures, or outsourcing service level agreements.

NS-CP-7

In our Business Network, there are reliable mediators and/or authoritative network institutions, such as associations or network leaders, capable of equitably solving conflicts and punishing unfair behaviors.

References

Andreoni, J. and Miller, J. (1993). Rational cooperation in the finitely repeated prisoner´s dilemma: Experimental evidence. Econ J 103: 570-585.

Axelrod, R. and Dion, D. (1988). The further evolution of cooperation. Science 242:1385–1390.

Bolton, G. E., Katok, E. and Zwick, R. (1998). Dictator game giving: Rules of fairness versus acts of kindness. Int J Game Theory 27: 269-99.

Brandt H., Ohtsuki H, Iwasa Y., and Sigmund K. (2006). A Survey on Indirect Reciprocity. IIASA Interim Report, IR-06-065. IIASA Studies in Adaptive Dynamics, n.123.

Carroll, G.R. and Hannan M.T. (2000) The Demography of Corporations and Industries. Princeton, NJ: Princeton University Press.

Chen, H. and Chen, T. (2003). Governance structures in strategic alliances: transaction cost versus resource-based perspective. Journal of World Business, Spring 38 (1): 1.

Combs J.G., Ketchen D. (1999). Explaining Interfirm Cooperation and Performance. Toward a reconciliation of predictions from the resource-based view and organizational economics. Strategic Management Journal, 20:867-888.

Fehr, E., Fischbacher, U. (2003). The nature f human altruism. Nature 425: 785-791.

Fehr, E., Fischbacher, U., and Gachter, S. (2002). Strong reciprocity, human cooperation, and the enforcement of social norms. Hum Nature-Int Bios 13: 1-25

Fink, M. and Kessler, A. (2010). Cooperation, Trust and Performance. Empirical Results from Three Countries. British Journal of Management, 21 (2): 469-483.

Gintis H. (2000). Game theory evolving: A problem-centered introduction to modeling strategic behavior. Princeton University Press.

Gintis, H., Smith, E. A. and Bowles, S. (2001). Costly signaling and cooperation. Journal of Theor. Biology 213: 103-119.

Greenhalgh, L. (2001). Managing strategic relationships: The key to business success. Free Press.

Hannan, M. T., & Freeman, J. (1977). The population ecology of organizations. American journal of sociology, 929-964.

Hannan, M.T., L. Polos, and G R. Carroll (2007) Logics of Organization Theory: Audiences, Code, and Ecologies. Princeton: Princeton University Press.

Hauert C., Traulsen A., Brandt H., Nowak M., and Sigmund K (2007). The emergence of Altruistic Punishment: via freedom to Enforcement. IIASA Interim Report, IR-07-053. IIASA Studies in Adaptive Dynamics, n.137.

Hillman, A. J., Withers, M. C. and B. J. Collins (2009). Resource dependence theory: A review. Journal of Management 35: 1404-1427

Hofbauer J., and Sigmund K. (2003). Evolutionary Game Dybamics. IIASA Interim Report, IR-03-078. IIASA Studies in Adaptive Dynamics, n.76.

Kryazhimskiy A.V.,Kleimenov A.F. (1998). Normal Behavior, Altruism and Aggression in Cooperative Game Dynamics. IIASA Interim Report IR-98-076.

Lavie, D. (2006). The competitive advantage of interconnected firms: An extension of the resource-based view. Academy of Management Review, Vol. 31, pp. 638–658.

Lieberman E., Hauert C., and Nowak M.A.(2005). Evolutionary dynamics on graphs. Nature 433: 312-316

Nowak M.A. (2006), Five Rules for the Evolution of Cooperation. Science, 314, 8: 1560-1563.

Nowak M.A. and Sigmund K. (1993). A strategy of win-stay, lose-shift that outperforms tit-for-tat in the Prisoner's dilemma game. In Nature 364: 56-58.

Nowak M.A. and Sigmund K. (2005). Evolution of indirect reciprocity.

Nowak M.A., Page K.M. and Sigmund K. (2000). Fairness Versus Reason in the Ultimatum Game. IIASA Interim Report, IR-00-057. IIASA Studies in Adaptive Dynamics, n.50.

Nowak MA and K Sigmund (1992). Tit for tat in heterogeneous populations. Nature 355: 250-253

Ohtsuki H., C HauertC., Lieberman E., and Nowak M.A. (2006). A simple rule for the evolution of cooperation on graphs and social networks. Nature 441: 502-505

Ostrom E. and Walker J. (2005). Trust and reciprocity: Interdisciplinary Lessons for Experimental Research. Russel-Sage Foundation.IIASA Interim Report, IR-05-079. IIASA Studies in Adaptive Dynamics, n.111.

Pfeffer, J., G. R. Salancik (1978). The External Control of Organizations: A Resource Dependence Perspective. New York, NY, Harper and Row.

Rainey P.B. and Rainey K. (2003). Evolution of cooperation and confict in experimental bacterial populations. Nature, Vol. 425, No. 6953. (4 September 2003): 72-74.

Ricciardi F., Cantù C. (2011). The Role of Altruism in Inter-firm Relationships: Long-term Value Creation in Business Networks. Proceedings of the 27th IMP-conference in Glasgow, 2011. Available online at: http://www.impgroup.org/uploads/papers/7754.pdf .

Roberts G. (1998). Competitive Altruism: from Reciprocity to the Handicap principle. Proceedings of the Royal Society, London

Sigmund K. (1998). Complex Adaptive Systems and the Evolution of Reciprocation. IIASA Interim Report, IR-98-100. IIASA Studies in Adaptive Dynamics, n.32.

Sigmund K. (2002). The Economics of Fairness. IIASA interim report, IR-02-020.

Sigmund K. (2007). Punish or Perish? Retaliation and Collaboration Among Humans. IIASA Interim Report, IR-07-054. IIASA Studies in Adaptive Dynamics, n.138.

Simon H. (1984), "On the Behavioral and Rational Foundations of Economic Dynamics,"Journal of Economic Behavior and Organization, 5(1): 35–56.

Trivers R.L. (1971). The evolution of Reciprocal Altruism. The Quarterly Review of Biology, 46 (1): 35-57.

Ulrich, D. The population perspective: review, critique, and relevance. Human Relations 40.3 (1987): 137-151.

West S.A., Griffin A.S., and Gardner A. (2007). Evolutionary Explanations for Cooperation. Current Biology, 17: 661-672

Williamson O.E. (2009). Transaction Cost Economics: The Natural progression. Nobel Prize Lecture,

Young H. P. (1996). Social coordination and social change. IIASA working paper 96-32.

4. Ecology of Innovation in Organizational Settings

Abstract Management and organization studies often ambiguous in distinguishing "creativity" from "innovation". This seems to have somehow hindered both a meaningful definition of the phenomenon of creativity, and a clear understanding of the relationship between creativity and its main desired outcome, i.e. successful innovation. Moreover, whilst numerous antecedents of creativity/innovation were identified by literature, they often looked contradictory: for example, anxiety was found positively correlated to creativity/innovation, but also relaxation did. The theories of creativity cited in literature were incapable of explaining this plethora of seemingly paradoxical outcomes. Then, in the second part of the Chapter, an integrated new theory of creative innovation is presented. This theory has been developed on the basis of some seminal writings on evolutionary dynamics. According to this theoretical view, successful innovation is made possible by refined homeostatic strategies that alternate opposing factors such as, for example, conformism and rebellion, optimism and pessimism, control and autonomy, rich resources and resource shortage, strong and loose network bonds. In other words, according to such view, creative innovation may be described as part of a "tacking path", similar to that of a sailing boat zigzagging against the wind. Managerial implications of this model are presented, along with two novel Scales consistent with it, aimed at measuring (i) the "Strategic Support to Innovation" in the organization, and (ii) its "Innovation Capabilities".

4.1 Introduction

In this study, creativity is understood as a characteristic of the processes through which subjects seek to solve their problems and/or to enrich their resource base for addressing possibly different problems in the future. The less a process of this type is repetitive and/or mechanically predetermined, the more it is creative. Differently from many authors (e.g. Amabile et al. 1996) I chose not to include the usefulness (or other measures of performance) of the outcome of the creative process in the definition of creativity, since I think that risk and even high failure rates are intrinsically embedded in the nature of creativity. And, after all, Leonardo Da Vinci's projects on flying machines were highly creative, even if they remained useless for centuries.

According to this definition, problem solving and/or resource grasping processes can be more or less creative, depending on whether their distance from the usual paths of problem solving and/or resource grasping is wide or narrow, independently from the perceived quality or innovativeness of their outcomes.

Management literature has studied creativity focusing on different types of processes (e.g., the product development process) and on different types of subjects performing creative processes: the unit of analysis (i.e. the subject whose capability of engaging in creative processes is under investigation) may be, for example, the individual person, the work team,

37

the organization, or the network. Different levels of creativity may reciprocally influence each other: for example, creativity at the individual level may influence creativity at the organizational level, but also vice versa (Bharadwaj and Menon, 2000).

In management literature, the concept of creativity is strongly linked to the concept of innovation. Some authors seem to imply that the association between creativity and innovation is so strong (e.g. Khanwalla and Mehta, 2004) that innovation can be considered as a proxy for creativity. But other authors describe or imply a more complex relationship between creativity and innovation: in fact, creative processes may result in dead ends, or in unsuccessful innovation (Amabile et al. 1996); on the other hand, in many cases the subjects innovate without previous engagement in strongly creative processes, for example by effectively just "copying and pasting" solutions developed elsewhere (Kao, 1990). In other words, there may be innovation without creativity, and there may be creativity without innovation.

The correlation between creativity and its main desired outcome, i.e., successful innovation, is described in many ways in literature, as the paragraph below will show more thoroughly. In some cases, management researches describe a simple, positive, linear one-way cause-effect relationship linking creativity to innovation; in other cases, more complex feedback effects between creative behaviors and successful innovation are hypothesized or implied. Other papers imply that the relationship may be more effectively described by an inverted U shape: after a certain limit, "too much creativity" may result in lower innovation performances. Other authors analyze the relationship between specific types or levels of creativity and specific types or levels of innovation: for example, product innovation tends to be seen as more linearly linked to individual creativity than, say, process innovation (Waples and Friedrich, 2011).

Another core issue in research on organizational creativity is the investigation on the antecedents of creativity for the different possible subjects (e.g. people, teams, organizations, networks) and for the different types of processes.

Again, the outcomes presented by literature are challenging, as it will be better detailed in the paragraphs below. In fact, some papers say that rewards are beneficial to creativity, whilst others say that they are detrimental; some papers say that a relaxed environment is fundamental, whilst others highlight the positive role of anxiety; some state that autonomy is necessary, but others claim that structured rules must be present; some stress the role of strong network bonds, whilst others demonstrate the importance of loose bonds, and so on.

The literature survey, in other words, yields a fragmented picture, both on the side of antecedents and on the side of consequences of creativity in organizational settings.

Attempts to build a comprehensive and consistent theoretical framework for understanding creativity and innovation, capable to explain the seemingly contradictory outcomes of the numerous empirical researches published so far, are very rare. This study seeks to provide an exploratory contribution to this goal.

4.2 Literature Search Method

The first purpose of this work was to assess whether the rich literature describing field research on organizational creativity provides outcomes which are overall consistent with some general explanatory theory of creativity in organizational settings.

Given this specific purpose, the literature search was designed on the basis of the method proposed by Vom Brocke et al. (2009).

The criteria for literature search were the following:

1. A representative sample of writings focusing on the antecedents of creativity in organizational settings and on the relationship between creativity and innovation was selected through database search;
2. Writings identified in step 1 were carefully analyzed and classified, and their main hypotheses or propositions on creativity were extracted;
3. Selective backward search was conducted from the richest and the most recent among the writings identified in step 1, in order to identify their most important theoretical references;
4. Selective forward search from the writings identified in step 3 was conducted, in order to find further, more recent writings that assess or represent the state-of-the-art in studies on organizational creativity;
5. Papers identified in steps 3 and 4 carefully analyzed and classified, and their main hypotheses or propositions on creativity were extracted.

The representative sample of writings (see step 1 of the list above) was selected as follows: the Econlit and Business Source Première databases were explored through the EBSCO search engine, using the keywords "creativity AND organization" for full text search throughout the Abstracts, in November 2011 and then again in May 2012 (to check for new entries). The search yielded 755 results overall. These writings were then selected through Abstract analysis, on the basis of the following question: does the paper focus on the antecedents of creativity and/or on the relationship between creativity and innovation in organizational settings?

The writings deemed compliant with this criterion were 57.

According to step 2 in the literature search method, these papers were then analyzed and classified in a Concept Matrix where each retrieved writing was structured in a record including the writing's specific answers, if present, to the following questions:

According to the writing under examination,

- What is creativity in organizational settings?
- What are the antecedents of creativity in organizational settings?
- What are the possible relationships between creativity and innovation in organizational settings?
- How can we consistently explain the key features of creativity?
- How can we explain the relationships between creativity and all its antecedents?
- How can we explain the relationship between creativity and successful innovation in organizational settings?

An overall survey of the Concept Matrix, once filled with data, allowed the identification of those writings where attempts to identify theoretical explanations of creativity-related phenomena were made.

Such writings were thoroughly analyzed both in their texts and in their Reference Lists, in order to conduct backward search aimed at the identification of further texts perceived as seminal by our disciplinary community as for the build-up of theoretical explanations for creativity-related phenomena (see step 3 in the list *Criteria for Literature Search*).

As a consequence, 8 further papers were added to the list of relevant writings and included in the Concept Matrix. Those paper were examined, and a selective Forward Search was conducted on their basis (see step 4 in the list *Criteria for Literature Search*), using the Google Scholar engine, to identify further, very recent papers (published not before 2005) whose Abstract showed specific focus on state-of-the-art analysis and/or general theoretical frameworks (and not on single, specific cause-effect relationships). This step yielded 4 further papers, which were included in the Concept Matrix.

After this process of literature search was concluded, a total of 57+8+4= 69 writings had been identified. This pool, though clearly far from being exhaustive, was deemed a proper sample to provide a first overview in order to answer the question triggering this literature search: *does the literature describing field research on organizational creativity provide outcomes which are overall consistent with some general explanatory theory on creativity-enabled innovation in organizational settings?*

4.3 Literature Survey

In the following Paragraphs, I will seek to synthesize the most interesting answers provided by the papers selected through the literature search process to the main research sub-questions:

1. What is creativity in organizational settings?
2. What are the antecedents of creativity in organizational settings?
3. What are the possible relationships between creativity and innovation in organizational settings?
4. How can we consistently explain the relationships between creativity and all its antecedents, and the relationship between creativity and successful innovation in organizational settings?

Following this framework, a structured synthesis of and comments on the outcomes of the literature survey will be provided in the following sub-paragraphs.

4.3.1 What Is Creativity in Organizational Settings?

My general impression, after the analysis of the retrieved writings, is that in several writings there is some difficulty in defining creativity in a way that is suitable for all creativity-related phenomena in organizational settings. Several papers leave their definition of this concept implicit, or use proxies or "is function of" definitions.

In any case, all of the writings say or imply that creativity is a process of creation of something. But what makes a process of creation a *creative* one? Some authors say or imply that it is the outcome's novelty and/or usefulness and/or brilliance that makes the difference, while other authors (a minority) focus on the very nature of the creation process in itself, independently from its performances.

A classical approach to creativity is that of Amabile et al. (1996), who see creativity as a possible capability of individual persons: "we define creativity as the production of new and useful ideas in any domain". This performance-oriented definition, rooted in psychological studies dating back to the 1950s, is the one that seems most commonly accepted or implied in management studies, according to the collected sample of writings. Some authors complement this definition with a wider list of possible outcomes of creativity: Woodman, Sawyer and Griffin (1993), for example, say that organizational creativity is "the creation of a valuable, useful new product, service, idea, procedure, or process". The problem with these definitions is that they imply that creation processes are not understandable as creative if their outcomes are not perceived as useful: as anticipated before, this type of definition would prevent us from considering consider creative all those projects by Leonardo Da Vinci that remained useless for centuries!

A restricted number of authors build on a more process-oriented definition of creativity: they are not so interested in the performances of the "ideas" that came up, but they rather focus on the specific features of the creation process in itself. As an example of this approach to the definition of creativity, I will quote George and Zhou (2007): "Creativity in the workplace (particularly on jobs that are not 'creative' in terms of the content of work tasks) often entails recognizing problems with the status quo and the need for change (Zhou and George, 2001); it also entails hard work, sustained effort over time, and persistence. It often necessitates a rejection of pre-existing schemas and a more bottom-up search for better ways of doing things and new ideas".

The reader will easily notice that, in this case, creativity is defined through the description of "what happens" during creative processes, independently from the nature and the successfulness of the outcome; moreover, according to this definition, moods and emotional attitudes such as rebellion and dissatisfaction, on the one side, and persistence and optimism, on the other side, are core factors of creativity. This definition of creativity may be more difficult to operationalize than the performance-oriented one by Amabile et al. (1996) described above, but it allows, in my opinion, a sounder distinction between creativity and innovation, and then a more thorough understanding of these complex phenomena.

These different definitions of creativity correspond to different scales to measure it as a variable or proxy. Amabile et al. (2006) assess creativity through a 6+6 items scale that measures the degree to which workers perceive that their own organizational area performs well in terms of innovation and productivity. In other words, and maybe quite oddly, Amabile et al.'s approach understands creativity as a performance, but this performance is measured by the declared opinion of the performers themselves - which sounds as a risky proxy, at least from a managerial point of view.

The scales chosen by Khanwalla and Mehta (2004) appear more consistent with the behavioral approach, in which creativity and its desired outcome, i.e., successful innovation, tend to be unified in a single concept. In their paper, innovational success is considered as an obvious proxy for creativity and it is measured by an aggregate of scales, including items assessing (i) the organization's image as for innovativeness; (ii) the extent to which current revenues were generated by recent product innovation; (iii) the successfulness of recent technological process innovations, measured on the basis of organizational records; (iv) the successfulness of recent operations-related innovations, measured on the basis of organizational records; (v) the successfulness of recent innovations in strategy, structure, and managerial practices, measured on the basis of organizational records.

The scale chosen by George and Zhou (2007), on the other hand, is more consistent with the process-oriented view of creativity implied by that paper. In that case, creativity is assessed through a 13 item scale measuring the opinion of a supervisor who is very familiar with the actual behaviors of the assessed worker. Examples of items are: "(the assessed worker) suggests new ways of performing work tasks" or "(the assessed worker) comes up with new and practical ideas to improve performance".

As the reader can see, different types of proxies imply even very different measures, and then concepts, of creativity and of innovation capabilities; this fact should be taken into account while reading the following paragraphs, where many different hypotheses involving so different proxies of creativity are compared and contrasted.

4.3.2 What Are the Antecedents of Creativity in Organizational Settings?

The list of the antecedents of creativity in organizational settings identified in literature is vast, fragmented and in many cases even paradoxical.

The antecedents identified during the literature survey were classified into nine groups, and namely: (1) environmental pressure; (2) corporate strategies and organizational structure; (3) supportive innovation management practices; (4) organizational climate; (5) network of relationships; (6) ethical factors; (7) expertise; (8) cognitive style; (9) personality traits and moods.

Environmental pressure is sometimes seen as a key factor for triggering creativity. For example, Kristensen (2006) in a single case study found that innovativeness and creativity are related to the contact with customers: in other words, the more a work group is exposed to direct contact with customers, the more it will be innovation-oriented. Vermeulen, Puranam and Gulati (2010) are concerned with the "corporate cholesterol" that tends to stiffen the organization when everything is going all right, the environment is not particularly challenging and the company becomes too stable. Khanwalla and Mehta (2004) hypothesized that environmental pressure is a key antecedent of creativity and innovation, and measured this variable with four scales (growing turbulence of output markets, growing turbulence of input markets, growing sophistication of customers, growing vulnerability to hostile acts). But they were surprised to find that, in their sample, there was no clear correlation between environmental pressure and creativity/innovativeness: some organization reacted to higher environmental pressure by enhancing innovativeness, whilst others preferred the status quo. Even more, several organizations displayed high innovative performances after periods of low environmental pressure.

In other words, a direct influence of environmental pressure on creativity is far from being demonstrated: further research is needed. A possible direction of investigation may be the hypothesis that (perceived, possible) environmental pressure does positively influence creativity, but this relationship is strongly mediated or moderated by other factors that may change such influence from positive to negative. What these possible moderators are, and under what conditions enter the playground, is a potentially interesting research question.

Corporate strategies were identified by Khanwalla and Mehta (2004) as auxiliary predictors of creativity and innovational success. Innovation-oriented corporate strategies were measured by scales assessing, for example, top management commitment in diversifying products and in entering new markets. The regression analysis confirmed the relationship, but the authors highlighted that innovation-oriented corporate strategies were weaker predictors of innovational success than, for example, innovation-oriented management practices. In particular, the correlation between innovation-oriented efforts in *organizational design* (i.e. changes towards higher administrative flexibility, hierarchic structure flatness, decentralization, and matrix structure with dual responsibilities) and innovational success was found particularly weak. This study, in other words, suggests a somehow bottom-up nature of the phenomenon of successful innovation, which seems difficult to influence through classical top management leverages, such as corporate strategies and general organizational design.

Supportive innovation management practices, on the other hand, are identified as very strong predictors of creativity by many authors. Practices originating in managerial decisions and directly affecting the processes that are expected to be creative are included in this group. The literature has built several, but sometimes opposing hypotheses in this stream of studies. For example, several authors insist that administrative control is detrimental to creativity (e.g. Desai, 2010) whilst other authors claim that "extolling corporate creativity at the expense of conformity may […] reduce the creative animation of business" (Levitt, 2002). Some researches corroborate the idea that evaluation, incentives and rewarding are beneficial to creativity (e.g. Cravens, Oliver and Stewart, 2010; Rosa, Qualls and Fuentes, 2008; Waples and Friedrich, 2011) whilst other studies demonstrate that there are situations in which evaluations, incentives and rewarding are detrimental (Eisenberg, 1999; Eisenberger and Rhoades, 2001). Many writings say or imply that procedures and process management are the "death kiss" for creativity (Grobman, 2005) and that procedures can cause paralysis within organizations (Kobe, 2010); but other authors, as for example Adler, Nguyen and Schwerer (1996) say that standardization does not kill creativity: on the contrary, systematic processes and clear organizational rules and structures were found positively correlated to creativity in product development, thanks to the mediating effect of trust (Brattstrom, Lofsten and Richter, 2012), and in many situation the stability provided by well-established rules is likely to support otherwise impossible innovation (Levitt, 2002). Some authors highlight the importance of providing sufficient resources for the project that is expected to be usefully creative (Amabile, 1996) but other authors found that generous resource allowance is not a reliable predictor of successful innovation (Weill, 1992). Some authors hypothesize that it is very important that workers are challenged and their targets are stretched in order to generate a certain amount of anxiety (Parker and Smith, 2011) that stimulates them to creative behaviors (Grobman, 2005; Khanwalla and Mehta, 2004) whilst others state that allowing for relaxation niches and serendipity is essential for new ideas to emerge (Ciborra, 2002).

The examples of the somehow contradictory nature of the research outcomes on management practices as predictors of creativity may go on. Among the selected writings, I could not find a single paper addressing this paradoxical state-of-the-art.

Organizational climate and culture defines a group of constructs that many authors identify as very strong predictors of creativity in organizational settings.

These constructs measures phenomena such as, for example, fair, constructive judgment of ideas, shared vision, fluid knowledge sharing, recognition for creative work, but also the presence of supervisors who serve as good work models and show confidence in the work group (Amabile et al., 1996). Khanwalla and Mehta (2004) propose further scales for measuring other aspects connected with organizational culture, such as, for example, the managerial entrepreneurship, the attitude to bend or bypass obstructing rules, the diffusion of participative decision-making, the interactive evolving of decisions, the space allowed for operating autonomy, the preference for external professional surveys, the level of interactivity of the managers with customers, suppliers and competitors. All these factors were found strongly correlated to innovational success. Other aspects of organizational culture that were found correlated with creativity and innovation are: encouragement of risk taking (McLean, 2005); organizational pride (Gouthier and Rhein, 2011); encouragement of non-complacent attitudes (Rosa, Qualls and Fuentes, 2008); some key characteristics of the physical work environment (Kao, 1990); humor, since it enables to manage stress and uncertainty and to identify with group culture (Heiss and Carmack, 2012) and it enables to see the world in a more relaxed, joyful and balanced perspective (Nargunde, 2011).

Also obstacles to creativity rooted in organizational culture or climate have been identified, such as, for example, the overmuch role of power games (Kleiner, 2003), political problems, or destructive internal competition (Amabile et al., 1996).

The (inter-organizational) *network of relationships* influences creativity, too, according to an emergent stream of studies. This stream of studies highlights the fact that many essential resources for organizational creativity often lie far beyond the organization's boundaries, and then the interaction with customers, suppliers, stakeholders or competitors may result in creative processes that could not have occurred in isolation.

For example, Steiner (2009) introduces the concept of "open creativity", similar to that of "open innovation" occurring within networks; whilst Maqsood, Walker and Finegan (2007) highlight the positive influence of "learning chains" involving supply chain partners, since learning chains facilitate the extension of knowledge advantage throughout the supply chain. Free flow of ideas and knowledge sharing are identified as key measures for innovation supportiveness of inter-organizational networks. An emergent stream of studies focuses on the relationships between network topology and creativity (Uzzi and Spiro, 2005), whilst the IMP group is providing a growing corpus of research on the influence of network bonds and relational interactions on innovation, value creation, and competitive advantage (Håkansson et al. 2004)

But, again, a more thorough examination of texts shows that the identified relationships between antecedents and creativity are not linear and pose interesting challenges to theory. For example, Håkansson et al. (2004) strongly believe that long-term network bonds are beneficial for innovativeness, but they must admit that there are cases in which long-term network bonds paralyze the actors in detrimental opportunism and conformism. Uzzi and Spiro (2005) notice that the relationship between the network characteristic they focus on and creativity is an inverted-U curve, with a threshold beyond which the specific network characteristics, that had proven so beneficial before, become detrimental. Also Perry-Smith and Shalley (2003) found a non-linear relationship between network position (and then the number and strength of network ties) and creativity: they then propose "a spiraling model, capturing […] the cyclical relationship between creativity and network position". Theoretical explanations to such complex cause-effect relationships are hardly attempted.

Ethical factors, too, are more and more taken into consideration as antecedents of creativity. Valentine et al. (2011), after a quantitative study, found that group creativity and corporate ethical values were positively related. Schepers and Berg (2007) stated that there is a relationship between knowledge sharing, procedural justice and work-environment creativity. Khazanchi and Masterson (2011) investigated the role of fairness in organizational creativity. But again, the more this emergent stream of studies deepens its investigations on the relationships between creativity and its antecedents, the more some of such relationships prove complex, non- linear and challenging from a theoretical point of view. The paper by Mueller and Kamdar (2001), addressing the paradoxical effects of reciprocation in the relationship between collaborative help and creativity, is a good example.

Expertise on the specific issue where innovation is expected is often mentioned as another key resource to fuel creativity (e.g. Amabile et al., 1996; Ciborra, 2002); nevertheless, again, it seems that also this relationship is not linear, since literature recognizes also the importance of factors which are the opposite of expertise,. i.e. ignorance, ingenuity and "outsiderness", as antecedents of creativity (Lorenz, 1973).

Cognitive styles are proposed as antecedents of creativity by the conceptual-skills theory of creativity. Boone and Hollingsworth (1990) state that creativity stems from cognitive skills and techniques, such as unconventional thinking, visualization of thoughts or models, reassembly and recombination of old learning. As a consequence, creativity training aimed at developing such skills in workers in order to enhance organizational creativity is suggested (Tan, 1998). Nagasundaram and Bostrom (1994) propose that such skills may be supported by Group Support Systems (GSS) in team based organizations. Nevertheless, Birdi (2005) assessed the effectiveness of creativity training programs and found that employees subject to such programs may display moderate improvements in creativity, but such positive effects are limited by possible unfavorable climate for innovation.

Personal traits and moods are at the core of the Attribute theory of creativity (Gundry, Kickul and Prather, 1994). According to this approach, creativity occurs where creative people are, and people is creative in that display personal characteristics such as openness, independence, autonomy, intuitiveness, and spontaneity (Ray, 1987). But also in this case, opposing voices raise in literature: for example, Levitt (2002) found that such "creative people" may be very dangerous in business settings if left free to carry on the innovation processes their way: they often "put forward irresponsibly" their ideas, they often prove incapable to walk through the last mile of the creative process, and their approach can lead to disaster. Also in this field, then, the relationship between creativity and its antecedents seems to be non-linear. This impression is confirmed by George and Zhou (2007) who found that, in a creativity supportive context, both strongly positive and strongly negative moods are antecedents of creativity. The work by George and Zhou (2007) is perhaps the unique case where the author provides the reader with a satisfactory explanation of the seemingly paradoxical nature of the antecedent-outcome relationship identified. As the author explains, according to the mood-as-information theory (Schwarz, 2002), moods have powerful effects on cognitive processes and behaviors. "That is, in order to adapt to the environment and function effectively, people's thought processes and behaviors are tuned to the information provided by their moods". Positive moods inform us that all is going well and then promote compliance with top-down strategies, fluid ideation, heuristic problem solving; on the other hand, negative moods signal a problematic state of affairs and propel us to figure out what's wrong, so encouraging dissatisfaction with the ideas at hand and a bottom-up, analytical approach. George and Zhou demonstrate that both groups of attitudes and behaviors are necessary for good creative performances, and show how the experience of both moods, which alternate at the individual and at the group level, lead to high creative performances. In conclusion, this paper proposes the theory of a "dual tuning" of creativity, on the part of the pair of opposing antecedents, which alternate in triggering complementary cognitive attitudes and behaviors.

4.3.3 What Are the Possible Relationships Between Creativity and Innovation in Organizational Settings?

Surprisingly, I found that the relationship between creativity and its main expected outcome, i.e. (successful) innovation, has not been thoroughly investigated so far.

As detailed above, authors often use (successful) innovation as a proxy for creativity, so taking for granted that the relationships between these two constructs should be considered strong and linear; but in the sample of literature I selected, I found no quantitative study specifically assessing the causal relationship between creativity and innovation, so it is hard to determine where this confidence in a strong and linear relationship between these two variables comes from.

On the other side, an underground concern emerges in many writings, about the possibility that just those antecedents, which proved so beneficial for coming out with new ideas, could be detrimental for successfully implementing such ideas. For example, Adler, Nguyen and Schwerer (1996) highlight the importance of standardization, control and process management when it comes to proper product development activities; and Levitt (2002) insists that creativity is not good per se, since it suffers from a sort of endemic sense of irresponsibility that can hinder, rather than help, the organization. If this is true, then it is logical to expect that creativity and successful innovation, instead of co-vary, may be antagonists to some extent, at least under certain conditions. Nevertheless, these important issues have been hidden in literature so far, under the ambiguity of a frequent fusion between the concepts of creativity and innovativeness.

4.3.4 How Can We Consistently Explain Creativity and Successful Innovation in Organizational Settings?

As the reader may have noticed, the antecedents of creativity/innovation capabilities identified above often tend to polarize into pairs of opposite constructs: for example, creativity or innovation capabilities are enhanced by freedom, but also by rules; by anxiety, but also by relaxation; by strong network bonds, but also by loose network bonds; by reciprocation, but also by opportunism; by independence, but also by feelings of belonging; by rich resources, but also by resource shortage; by positive moods, but also by negative moods; and so on.

This paradoxical framework emerges only from an overview of the whole selected literature: individual writings tend to avoid these awkward contradictions by concentrating on the effect of a single antecedent, and not on the effect of its opposite. Also writings that concentrate on a pair of opposite antecedents, like for example that on the role of positive and negative moods (George and Zhou, 2007), do not address the fact that the polarization into pairs of opposite antecedents seems to be the background sound-track of the phenomenon of creativity as a whole.

Since these contradictions are kept hidden, or at least overlooked, by authors, who tend to concentrate on one side at a time of these complex cause-effect relationships, it is not surprising that I could not find in the selected literature any consistent overall theoretical explanation, capable to provide a synthetic view of the phenomenon of creative innovation, and to explain all the seemingly contradictory outcomes of field researches.

Nevertheless, the examined literature includes some very useful hints, in order to figure out how a good explanatory theory of creative innovation should look like:

- Many among the proposed antecedents may be affected by a threshold effect: for example, anxiety may be beneficial to creativity, to some extent; but beyond a certain amount it may become detrimental;
- Many of the opposing antecedents described may co-exist because they do not display contemporaneously: for example, positive and negative moods alternate like in an oscillatory phenomenon, so that their respective and complementary beneficial effects can unfold (and their respective detrimental effect can be kept under control), in turns.

As a consequence, research on creativity could benefit from a theoretical framework capable to explain how, and why, threshold effects and/or oscillatory phenomena affect creativity.

4.4 An Evolutionary Approach to Creativity and Innovation

There is a branch of science that is particularly used to understand and manage threshold effects and oscillatory phenomena: it is the study of adaptive dynamics in ecological settings (West et al., 2007).

Recent studies are blurring the boundaries between sciences of life, psychology and sociology, and provide us with valuable theories, which may be very useful to enrich management literature (Ricciardi, 2011).

The specific possible utility of the evolutionary approach in tackling organizational issues is that it can see organizational phenomena (even the most intangible, such as culture, emotions or social bonds) not as photos, but as frames from very long motion pictures. This powerful understanding of long-term historical dynamics enables the evolutionary research to produce theoretical tools and outcomes that would be otherwise hardly accessible.

For example, some studies on adaptive dynamics (e.g. West et al. 2007), soundly rooted in the evolutionary approach, are providing us with more and more powerful analyses on (organizational) altruism and cooperation, featuring an unprecedented understanding of the complex conditions under which altruistic cooperation is sustainable and effective, or not, and is then likely to develop, or to fade (Westerlund and Rajala, 2010). In other words, in the last decades the evolutionary thought has gone far beyond certain popular, although old-fashioned simplifications (such as the mere "survival of the strongest"), and has become capable to address highly complex socio-cultural phenomena, coming out with precise, interesting and often counter-intuitive explanations. Thus, even if the creativity issue is traditionally perceived as a prerogative of the humanities and of social sciences, this study proposes to borrow some possibly fertile theoretical outcomes also from the sciences of life.

In effect, *innovation processes* are at the very core of evolutionary studies. The basic strategy of evolution is about being able of an effective innovation at the right time. This study will focus on two aspects enlightened by evolutionary studies: (i) the risks of evolutionary irreversibility in innovation processes, and (ii) adaptive dynamic strategies to reduce such risks.

As for the risks of evolutionary irreversibility, which may always transform every innovation into a poisoned apple, the main reference is Stephen Jay Gould (1980); as for strategies capable of reducing the risks of evolutionary irreversibility, the main reference is Konrad Lorenz (1973). The next sub-paragraphs will be dedicated to very synthetically present Gould's and Lorenz's works.

4.4.1 The Main Evolutionary Problem with (Creative) Innovation: Irreversibility

Innovation (be it creative or not) is the core strategy of life, since it allows fitting; but it has a cost, and this cost is flexibility loss. Every innovation creates new internal constraints that will influence all further possible developments of the involved subjects. Stephen Jay Gould has thoroughly studied this phenomenon: in his famous writing on the panda's thumb (1980), he demonstrated that the panda developed a strange, not completely effective sixth finger from its sesamoid bone, since the previous evolutionary history of this animal had hindered the use of the first finger as an opposable finger. Gould proposes very interesting relationships between evolutionary processes and poor design, demonstrating that innovation irreversibility is at the very base of both evolutionary success and failure.

The evolutionary importance of creativity is that it allows short-term innovations, which can be stored in high-level store facilities, like an individual's memories or a community's school texts, without modifying the deepest and more unreachable layers of the subject's knowledge base (like the DNA). But, on the other side, creativity is a costly effort, and then its outcomes must be exploited and made efficient: that's why human beings have developed strong social and emotional impulses to store successful innovations' outcomes in habits, traditions, and maps. Conformism was evolutionarily developed to protect previous, precious creativity from oblivion, criticism and subversion.

Creative attitudes, then, although vital for our competitive strategies, need *antidotes* and *homeostatic suppressers* for two main reasons: (i) they may be unsuccessful, i.e. they may provide products or solutions that do not work; and (ii) they may be successful, which can be also a problem in the long run. If the creative behavior proves unsuccessful, the costs of the useless endeavors may result in severe consequences, possibly hindering or even preventing other, although necessary, efforts. If, on the contrary, the creative behavior proves successful, the subject (e.g. the organization) will be evolutionarily modified by its creative performances, and this may result in internal constraints with possible negative effects on further evolutionary processes (like "panda's thumbs"). This happens because each success, in the evolutionary dynamics, tends to result in habits and traditions, so threading certain paths (and making other paths inaccessible) for further innovations. In other words, from an evolutionary point of view, the "dark side" of creativity is irreversibility: irreversible waste of energies, in case of failure; irreversible loss of flexibility, in case of success.

Creativity has then a sometimes counterproductive, and often paradoxical, nature. How can it be managed? Konrad Lorenz (1973) provided a fascinating answer to this question in his studies on the natural history of knowledge.

4.4.2 The Homeostatic Mechanisms and the Tacking Nature of Innovation

According to Lorenz, learning and creative innovation are complex phenomena: to successfully manage them, it is impossible to rely on linear cause-effect relationships. Pairs of *opposing* factors, on the contrary, such as for example conformism and rebellion, can succeed in creating a dynamic ecology of innovation, on the basis of an homeostatic swinging around an invisible optimal course.

In this view, creativity is a particular way to create new solutions for actual or potential problems of any kind. Creativity implies the use of *linking*: during creative processes, things and concepts are linked in a novel way. For example, an analogy between two different situations is perceived, and then the solution adopted for the first situation is translated and adapted into the second situation, so creating a new solution for the problem at hand.

Two important learning processes are usually involved in creative activities and in novel concept linking: trial-error, and exploration. Trial-error concentrates on the solution of a specific problem which is perceived as interesting at that moment, whilst exploration is aimed at widening the knowledge base, even beyond the strict needs of the moment, "just to know". Exploration and trial-error may occur both in the real world, and in the imaginary world of abstract models, maps and representations.

Trial-error and exploration are usually triggered by the feeling that something new should be found. This feeling may stem both from positive (e.g. curiosity, ambition) and from negative (e.g. pessimism, rebellion) moods; both from resource and freedom abundance (e.g. relaxation, boredom) and from resource and freedom shortage (stress, pressures).

But trial-error and exploration cannot go on indefinitely: as explained before, since there are thresholds beyond which trial-error and exploration become unsustainable, in healthy systems there are feedbacks mechanisms that cyclically stop these processes, and drive energies towards the consolidation of what has been learned. Knowledge acquired through creative processes, in fact, is wasted unless it is stored safely, i.e. managed so that it is not forgotten and it can be re-used, also as a basis for further innovation.

Emotional and social triggers, then, lead to habits, traditions, procedures, rituals, myths: ways to store knowledge in deeper, safer layers, where oblivion and dispute are more difficult. As a consequence, what was once novel easily becomes customary or even a dogmatic constraint. The paradoxical aspect of creative behavior is that, the more it is successful, the more it tends to result in "anti-creative" standardization. Whilst in the creative phase the main learning strategies are trial-error and exploration, in the consolidation phase the main learning strategies are imitation and training.

In other words, both at the individual and at the social level, there is a cyclical alternation between creative and consolidation processes, between improvisation and control, between learning based on trial-error or exploration and learning based on imitation or training. These two phases are obviously opposing and they tend to create conflicts between different psychological attitudes, cognitive tools, organizational resources and rewards/punishment systems.

How can this paradoxical opposition be managed?

Lorenz resorts to the metaphor of the sailing boat. It is impossible for a sailing boat to course directly straight against the wind: the only way to go on against the wind is to slip-slide, alternating two opposing courses which, aggregated, result in the right direction.

Similarly, the wind of competition cannot be tackled with a single, consistent attitude: two opposing attitudes are necessary, and each competing unit (be it an individual or a business network) must periodically shift from one strategy to the other, before the course becomes irreversibly too far from fitness (Figure 4.1 synthetically illustrates this concept).

As a consequence, a competing subject remains viable as long as it is capable to frequently tack from creativity to consolidation, and vice versa. This oscillatory behavior is not necessarily planned or designed, since it may spontaneously stem from emotional attitudes and cognitive behaviors triggered by social and psychological homeostatic mechanisms.

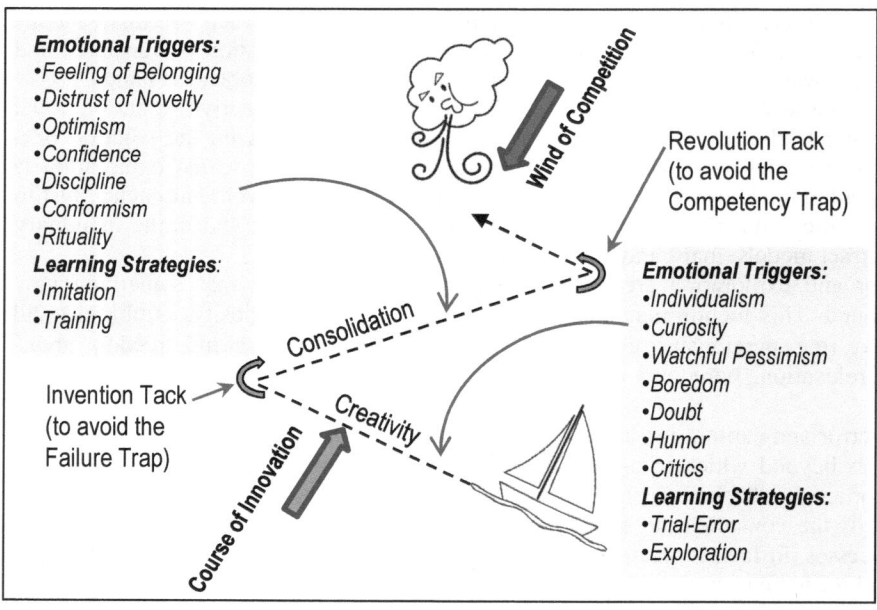

Fig. 4.1 The Tacking Strategy for Innovation. Elaborated from Lorenz (1973)

This is true both at the macro level (for example, a firm can alternate periods or even years of consolidation and periods or years of intense creative change) and at the micro level (many tasks are pursued by alternating short moments of "interstitial creativity" and short moments of partial consolidation). In healthy organizational settings such as, for example, business networks, firms, business units, or project teams, both creative and consolidation capabilities are simultaneously present, and are allowed to alternately set the course, so that the organizational boat can alternatively avoid both the "competency trap" (too much rigidities and lock-ins) and the "failure trap" (too many ineffective trials and explorations).

4.5 Measuring Organizational Innovation

I suggest that the understanding of creativity proposed above, stemming from the evolutionary approach, should be considered compatible with all the well-known and widely used management theories of creativity (e.g. the attribute theory, the conceptual-skills theory, the behavioral theory, see Gundry, Kickul and Prather,1994) and with all the creativity-related theories rooted in adaptive dynamics; moreover, it offers an integrated and unifying view on the phenomenon, providing a sound explanation to the otherwise awkward presence of numerous opposing antecedents to creativity and innovation.

According to the view proposed here, creative innovation may be described as a "tacking path", similar to that of a sailing boat zigzagging against the wind. I propose that this view may usefully complement the well-established theory of the recursive circle comprised of design, implementation and feedback. It is this paradoxical and sometimes counterproductive nature of creativity that makes the ecology of innovation so complex: organizational design cannot guarantee that organizational creativity will be successful, but can support those factors that make creative failure *and* success less dangerous. Only under these conditions can creativity be sustainable in the long run.

For a healthy ecology of creativity, it is important that the creative process is based on *dual tuning*, where complementary factors alternate in triggering complementary behaviors, which cyclically minimize each other's weaknesses and threats. In fact, *long-run* creativity is possible only if, in the deeper layers of the subject's knowledge base, strong rigid "a priori" procedures and beliefs are available to build upon – or to fight against. On this basis, the organization can successfully alternate conformism and rebellion, optimism and pessimism, feelings of belonging and individualism, tunnel visions and wide horizons, both at the individual and the at the social level (with a sort of emotional and attitudinal "division of labor").

According to this evolutionary view, then, the necessary antidotes to creativity are, in effects, antidotes to the undesired irreversible loss of potential and of flexibility that may result from both successful and unsuccessful creative processes. Anthropologically, such antidotes are provided by refined emotional and social mechanisms, such as rebellion, laziness, group rivalry, conservatism, boredom, or pessimism; but managerial literature has quite overlooked the surprising potential of such aspects so far.

In this paragraph, a definition of Organizational Innovation will be proposed, consistent with the theoretical view described in Paragraph 4.4. Then, the factors associated to high Innovation Capabilities in organizations will be discussed and two Scales will be built consistently, in order to measure these capabilities.

Organizational innovation is the outcome of change made

- within the organization (as for culture, structure, processes and/or systems),
- and/or in the organization's market and business network (as for customer base, relationships with suppliers, reputation, etc.)
- and/or in the organization's products (goods and services).

Innovation often implies a certain degree of creativity, i.e. it implies (see Paragraph 1.1), at least to some extent, non-repetitive, non-standardized problem solving and resource grasping, based on trial-error, exploration and novel concept linking. It is possible, indeed, that also non-creative innovation occurs (e.g. a procedure is mechanically changed in order to comply with a new norm; an IT solution is simply copied and pasted from a business unit to another), but this type of innovation is usually very easy to imitate, yields no long-term competitive advantage, and generally does not imply specific capabilities in addition to those necessary to perform successful creative innovation. Then, in the following investigation on the features that should be assessed to measure the innovation capabilities of the organizations, non-creative innovation processes will not be taken into consideration.

On the other hand, even if creativity is generally associated to all the most relevant innovation processes, there is not a linear relationship between creativity and successful innovation. As the model of Figure 4.1 predicts, if the course of the organization is set only by creativity (i.e. by trial-error and exploration, without consolidating the results) for a too long time, the organization will crash against the "failure trap", i.e. will fail the task of providing the organization with suitable, structured, deliverable, efficient new solutions for the problems at hand. It is important to notice that this is not a threshold effect, it is an oscillatory phenomenon. Good performances do not stem just from a static "balance" between creativity and consolidation; good performances stem from developing both creativity and consolidation capabilities, along with the capability to timely switch from creativity to consolidation attitudes, behaviors and strategies, and vice versa.

According to this theoretical framework, then, the organization's Innovation Capabilities are made up of four different capabilities, that should be developed both at the strategic level and at the operational level:

1. Creative Capabilities
2. Consolidation Capabilities
3. Invention Capabilities (i.e. capabilities to turn from creativity to consolidation)
4. Revolution Capabilities (i.e. capabilities to turn from consolidation to creativity).

At the operational level, creative capabilities tend to overlap with the revolution capabilities, they imply in both cases dissatisfaction with the present, and desire for something new. Similarly, at the operational level, consolidation capabilities tend to overlap with invention capabilities, since they imply in both cases expertise and get-the-job-done attitudes.

As a consequence, the construct "Organization's Innovation Capabilities" was developed in two dimensions:

1. the dimension "Capability of Novelty Generation", including revolution and creative capabilities, and
2. the dimension "Capability of Novelty Management", including invention and consolidation capabilities.

On the other hand, the discussions with the five experts involved in this research (see Chapter 1) confirmed that the presence of Innovation Capabilities is not enough: innovation oriented competitive strategies must also be developed by the top management team, to make the innovation tacking course possible in the long run and at the macro level, i.e. at the level of the whole organization or even at the inter-organizational level. Strategies are needed that protect, or at least do not hinder, the Capability of Novelty Generation, and particularly the capability of generating disruptive change, because this capability is the most subject to be destroyed in mature organizations.

Then, also the construct "Strategic Support to Innovation" was developed, utilizing not only the theoretical tools described so far, but also those developed by literature on exploration / exploitation, on organizational ambidexterity, and on dynamic capabilities.

This approach is focused on top-down strategic management and tends to overlook the importance of bottom-up, interstitial or spontaneous innovation capabilities; on the other hand, this literature is very useful to build the construct "Strategic Support to Innovation" consistently with the theory presented here, because it stresses the importance of dual tuning and of long-term fitness. The publication by O'Reilly and Tushman (2008) was used to extract specific suggestions for developing the Construct and the related Scale.

4.5.1 Strategic Support to Innovation (SS)

8 Items Scale

Do you agree with the following statements? Please consider your organization's situation **in the last five years**. *(1=strongly disagree, 5=strongly agree, 3=neutral).*

SS-1
Our top management wants to position our organization as a unique one in its industry in the way it operates.
SS-2
Our top management is committed to diversify our products/activities and to enter new markets.
SS-3
Our organization strategically aims at pioneering new products/services.
SS-4
Our organization strategically aims at the highest quality and specialization of products/services.
SS-5
In our organization, intensely innovative periods alternate with consolidation periods, in which the achievements are rationalized and efficiency is consolidated.
SS-6
Our management periodically commissions professional market surveys, and/or SWOT-diagnostic studies, reorganization studies, morale surveys, customer satisfaction surveys, etc., to identify new opportunities and areas of innovation and improvement.

SS-7

Our organization has created at least one area, or department, or business unit, or spin-off or joint venture, that is quite protected from short-term cost/revenues pressures and has the mission to develop radically new capabilities or entering completely new markets.

SS-8

Our organization has developed privileged relationships with those external subjects (e.g. suppliers) that may allow us to develop radically new capabilities or entering completely new markets.

4.5.2 Organization's Innovation Capabilities (IC)

14+10 Items Scale

*Do you agree with the following statements? Please consider your organization's situation **in the last three years**. (1=strongly disagree, 5=strongly agree, 3=neutral).*

First Dimension: Capability of Novelty Generation (IC-NG)

IC-NG-1

Our managers use benchmarking to generate new and useful ideas.

IC-NG-2

Our management invests in the cooperation with customers, suppliers, competitors and/or research institutions for exchanging, securing or testing out innovative ideas.

IC-NG-3

Our organization strategically aims at offering personalized products/services to customers.

IC-NG-4

In our organization, there is at least one high-level and influential manager coming from other industries or from different professional experiences.

IC-NG-5

Our organization sometimes hires people coming from the best competitors.

IC-NG-6

In our organization, initiative and experimentation of new solutions are encouraged and rewarded.

IC-NG-7 (reverse item)

In our organization, complying with habits and procedures is more rewarding than finding new solutions.

IC-NG-8

Our organization's managers are very active in exploring the environment (e.g. by establishing new relationships, learning new things, visiting new markets), even if this is not directly aimed at a specific business goal.

IC-NG-9

Our organizations' policies include structured initiatives for personnel's education and re-training, and/or encourage the employees' participation in seminars, conferences, professional associations, cultural circles, etc.

IC-NG-10

In our organization, there are incentives for employees to point out inefficiencies and problems and to suggest new solutions and improvements.

IC-NG-11 (reverse item)

In our organization, people who are obsequious with their bosses and unfair with colleagues are more likely to advance their careers.

IC-NG-12

In our organization, those who take decisions can easily access reliable and updated information about customers' feedbacks, stakeholders' expectations, suppliers' problems and potentialities, etc.

IC-NG-13

Bright, innovative young professionals are recurrently hired and given challenging assignments in our organization.

IC-NG-14

Our organization has activated effective cooperation or partnerships with other businesses, research institutions, etc., in order to achieve superior know-how or technological competences.

Second Dimension: Capability of Novelty Management (IC-NM)

IC-NM-1

In our organization, there are several very experienced people, who know our business and our industry very well.

IC-NM-2

Our organization is capable of effectively adopting successful solutions developed elsewhere, thanks to systematic benchmarking, updating initiatives, and/or market exploration.

IC-NM-3

Our management takes a more sophisticated approach than our competitors', and is more familiar with advanced IT solutions.

IC-NM-4

Our organization has proved capable of effective planning and project management.

IC-NM-5

Our organization has proved capable of effective auditing.

IC-NM-6

Our organization has proved capable of effective process management.

IC-NM-7

When an innovation is decided for in our organization (e.g. a new on-line sales channel), a multi-functional or multidisciplinary team or task force (e.g. sales+IT+warehouse managers) is usually established in order to manage the innovation process.

IC-NM-8

After being implemented, innovations are systematically and constructively monitored, in order to identify successes, problems and possible areas of improvement.

IC-NM-9

Our organization has a strong identity and a recognized tradition as for products/services, quality levels, and way of working.

IC-NM-10

Hard work is part of our organization's culture.

References

Amabile T.M., Conti R., Coon H., Lazenby J., Henon M. (1996). Assessing the work environment for creativity. The Academy of Management Journal, Vol. 39, Issue 5, Oct 96, pp. 1154-1184.

Adler S., Nguyen A.M., Schwerer E. (1996). Getting the Most out of Your Product Development Process. Harvard Business Review; Mar/Apr96, Vol. 74 Issue 2, p134-152.

Bharadwaj S., Menon A. (2000). Making Innovation Happen in Organizations: Individual Creativity Mechanisms, Organizational Creativity Mechanisms or Both? Journal of Product Innovation Management; Nov2000, Vol. 17 Issue 6, p424-434.

Birdi K.S. (2005). No idea? Evaluating the effectiveness of creativity training. Journal of European Industrial Training; 2005, Vol. 29 Issue 2, p102-111

Boone L.W., Hollingsworth A.t. (1990). Creative thinking in business organizations. Review of business, Fall 1990, Vol. 12, Issue 2.

Brattstrom A., Lofsten H., Richtner A. (2012). Creativity, Trust and Systematic Processes in Product Development. Research Policy, May 2012, v. 41, iss. 4, pp. 743-55

Ciborra C. (2002). The Labyrinths of Information. Challenging the Wisdom of Systems. Oxford University Press.

Cravens K.S., Oliver E.G., Stewart J.S. (2010). Can a positive approach to performance evaluation help accomplish your goals? Business Horizons; May2010, Vol. 53 Issue 3, p269-279.

Desai D.A. (2010). Co-creating learning: insights from complexity theory. Learning Organization; Aug2010, Vol. 17 Issue 5, p388-403.

Eisenberg J. (1999). How Individualism-Collectivism Moderates the Effects of Rewards on Creativity and Innovation: A Comparative Review of Practices in Japan and the US. Creativity & Innovation Management; Dec99, Vol. 8 Issue 4, p251.

Eisenberger R., Rhodes L. (2001). Incremental Effects of Reward on Creativity. Journal of Personality & Social Psychology; Oct2001, Vol. 81 Issue 4, p728-741.

George J.M., Zhou J.C. (2007). Dual Tuning in a Supportive Context: Joint Contributions of Positive Mood, Negative Mood, and Supervisory Behaviors to Employee Creativity. The Academy of Management Journal, vol. 50, no. 3, Jun 2007, pp. 605-622.

Gould, J. (1980). The panda's thumb. New York: Norton.

Gouthier M.H.J., Rhein M. (2011). Organizational pride and its positive effects on employee behavior. Journal of Service Management; 2011, Vol. 22 Issue 5, p633-649.

Gundry L.K., Kickul J.R., Prather C.W. (1994). Building the creative organization. Organizational Dynamics; Spring94, Vol. 22 Issue 4, p22-37.

Grobman B. (2005). Complexity theory: a new way to look at organizational change. Public Administration Quarterly; Fall/Winter2005, Vol. 29 Issue 3/4, p351-384

Håkansson H., Ford D.I., Gadde L., Snehota I., Waluszewski A. (2010). Business in networks. John Wiley & Sons.

Heiss S.N. and Carmack H.J. (2012). Knock, Knock; Who's There?: Making Sense of Organizational Entrance Through Humor. Management Communication Quarterly; Feb2012, Vol. 26 Issue 1, p106-132.

Kao J. (1990). The Entrepreneurial Organization. Prentice-Hal.

Khanwalla P.N., Mehta K. (2004). Design of Corporate Creativity. Vikalpa: The Journal for Decision Makers; Jan-Mar2004, Vol. 29 Issue 1, p13-28.

Khazanchi S., Masterson S.S. (2011). Who and what is fair matters: A multi-foci social exchange model of creativity. Journal of Organizational Behavior; Jan2011, Vol. 32 Issue 1, p86-106.

Kleiner A. (2003). Are You In with the In Crowd? Harvard Business Review; Jul2003, Vol. 81 Issue 7, p86-92.

Kobe C. (2010). Can structures foster creativity and innovation? The propositions based on a Giddens-inspired framework. International Journal of Product Development; 2010, Vol. 11 Issue 3, p166-176.

Kristensen K.S. (2006). Losing innovativeness: the challenge of being acquired. Management Decision; 2006, Vol. 44 Issue 9, p1161-1182.

Levitt T. (2002). Reactivity is not enough. Harvard Business Review; Aug2002, Vol. 80 Issue 8, p137-145.

Lorenz, K.(1973) Behind the Mirror. A search for a natural history of human knowledge, Harcourt Brace, New York .

McLean L.D. (2005). Organizational Culture's Influence on Creativity and Innovation: A Review of the Literature and Implications for Human Resource Development. Advances in Developing Human Resources; May2005, Vol. 7 Issue 2, p226-246.

Maqsood T., Walker D., Finegan A. (2007). Extending the "knowledge advantage": creating learning chains. Learning Organization; Mar2007, Vol. 14 Issue 2, p123-141.

Mueller J.S., Kamdar D. (2011). Why Seeking Help From Teammates Is a Blessing and a Curse: A Theory of Help Seeking and Individual Creativity in Team Contexts. Journal of Applied Psychology; Mar2011, Vol. 96 Issue 2, p263-276.

Nagasundaram M., Bostrom R.P. (1994). The Structuring of Creative Processes using GSS: A Framework for Research. Journal of Management Information Systems; Winter94/95, Vol. 11 Issue 3, p87-114.

Nargunde A.S. (2011). HR: Humour Resource. Indian Development Review, January-June 2011, v. 9, iss. 1, pp. 83-87.

O'Reilly III, C. A., Tushman, M. L. (2008). Ambidexterity as a dynamic capability: Resolving the innovator's dilemma. Research in Organizational Behavior, 28, 185-206.

Parker S., Smith M. (2011). The power of anxiety. Chief Learning Officer; Apr2011, Vol. 10 Issue 4, p27-51.

Perry-Smith J.E., Shalley C. (2003). The social side of creativity: A static and dynamic social network perspective. The Academy of Management Review, Vol 28(1), Jan 2003, 89-106.

Ray M. (1987). Strategies for Stimulating Personal Creativity. Human Resources Planning, Vol. 10.

Ricciardi F. (2011). Beyond Darwin: the Potential of Recent Eco-Evolutionary Research for Organization and Information Systems Studies. In Carugati A., Rossignoli C. (eds.), Emerging Themes in Information Systems and Organization Studies. Springer, Heidelberg.

Rosa J.A., Qualls W.T., Fuentes C. (2008). Involving mind, body, and friends: Management that engenders creativity. Journal of Business Research; Jun2008, Vol. 61 Issue 6, p631-639.

Schepers P., Berg P.T. (2007). Social Factors of Work-Environment Creativity. Journal of Business & Psychology; Spring2007, Vol. 21 Issue 3, p407-428.

Schwarz, N. 2002. Situated cognition and the wisdom of feelings: Cognitive tuning. In L. Feldman Barrett and P. Salovey (Eds.), *The wisdom in feelings*, 144-166. New York: Guilford.

Steiner G. (2009). The Concept of Open Creativity: Collaborative Creative Problem Solving for Innovation Generation - a Systems Approach. Journal of Business & Management; 2009, Vol. 15 Issue 1, p5-33.

Tan G. (1998). Managing Creativity in Organizations: a Total System Approach. Creativity & Innovation Management; Mar1998, Vol. 7 Issue 1, p23.

Vermeulen F., Puranam P., Gulati R. (2010). Change for change's sake. Harvard Business Review; Jun2010, Vol. 88 Issue 6, p70-76.

Uzzi B., Spiro J. (2005). Collaboration and Creativity. The Small World Problem. American Journal of Sociology, Vol. 111, no.2, sept. 2005.

Valentine S., Godkin L., Fleischman G., Kidwell R. (2011). Corporate Ethical Values, Group Creativity, Job Satisfaction and Turnover Intention: The Impact of Work Context on Work Response. Journal of Business Ethics; Feb2011, Vol. 98 Issue 3, p353-372.

Vom Brocke J., Simons A., Niehaves B., Riemer K., Plattfaut R., Cleven A. (2009). Reconstructing the Giant: On the Importance of Rigour in Documenting the Literature Search Process. Proceedings of the 17th European Conference on Information Systems (ECIS 2009).

Waples E.P., Friedrich T. (2011). Managing Creative Performance: Important Strategies for Leaders of Creative Efforts. Advances in Developing Human Resources; Jun2011, Vol. 13 Issue 3, p366-385.

Weill P. (1992). The relationship between investment in information technology and firm performance: a study of the valve manufacturing sector. Information systems research, December 1992, Vol. 3, no. 4, pp. 307-333.

West S.A., Griffin A.S., and Gardner A. (2007). Evolutionary Explanations for Cooperation. Current Biology, 17: 661-672.

Westerlund, M. & Rajala, R. (2010). Learning and innovation in inter-organizational network collaboration. Journal of Business & Industrial Marketing, 25(6): 435-442.

Woodman R.W., Sawyer J.E., Griffin R.W. (1993).Toward a Theory of Organizational Creativity. *The Academy of Management Review* Vol. 18, No. 2 (Apr., 1993), pp. 293-321.

Zhou, J. & George, J. M. 2001. When job dissatisfaction leads to creativity: Encouraging the expression of voice. Academy of Management Journal, 44: 682-696.

5. Motivations for Business Networking

Abstract Theories on business networks provide poor understanding of situations where it is the network itself, although featuring high degrees of loyalty, reciprocity, efficiency and trust, that seems to hinder innovativeness; on the other side, theories on business networks have not systematically investigated the different possible motivations that drive different business actors towards specific business networks. This Chapter suggests that these two gaps in theory are correlated. In fact, the three field researches presented here confirm that different business networks attract and select different actors and are effective in pursuing different goals, and these differences may strongly influence the innovation performances of the organizations. In the qualitative worst-practice research presented here, three networks where sustainable innovation processes were to some extent hindered or damaged or overlooked by the network itself were chosen as exemplary cases. After comparing and contrasting the outcomes of the case studies, the extant management literature, the suggestions from the theory of fairness and cooperation developed in Chapter 3 and from the theory of innovation developed in Chapter 4, the different dimensions of Network Effectiveness, corresponding to different "Motivations to Business Networking", are discussed and operationalized. Possible positive and negative influences of the different possible Motivations for Networking on the organization's innovational success are hypothesized.

5.1 Introduction

Whilst some business networks seem to constitute environments where viable, sustainable innovation is more likely, other business networks seem to discourage, or even hinder, (good) innovation processes.

Some networks, for example, behave as conservative lobbies, where actors actively cooperate to *prevent* any changes. Other networks do promote inter-organizational innovation, but this cooperation regrettably results in a more efficient destruction of common resources.

Literature on business networks is generally focused on positive features of network-enabled innovation processes (e.g. Björk and Magnusson, 2009; Westerlund and Rajala, 2010), and often fails in theoretically explaining situations where even well-structured networks, with high degrees of loyalty, reciprocity and mutual trust, seem to make things get worse, and good innovation get killed. Authors, of course, are aware of some possible counter-productive dynamics within business networks (e.g. Håkansson et al., 2009), but they often just touch on this issue, briefly describing the phenomenon as a sort of strange aberration, whose causes are not thoroughly investigated.

In this chapter, thus, three cases will be presented, in which the dynamic evolution of different networks resulted in conservatism, and/or in bad, non-sustainable, counter-productive innovation. I will seek to demonstrate that the qualitative analysis of such "worst practice" cases can provide interesting tips for business network theories.

In particular, the qualitative nature of the research enabled me to focus on the historical evolution of the analysed networks, and on the importance of the different *motivations* to business networking on the part of decision makers. The analysed cases, when compared and contrasted with the existing literature, led me to suspect possible correlations between (i) different motivations to business networking, (ii) attitudes to cooperation and to fairness within the network environment and (iii) different network-enabled innovation performances. I found that the concept "motivation to business networking" has been only partially investigated by literature so far, and consequently I sought to discuss this concept within a consistent theory of inter-organizational relationships, and to translate it into usable variables, suitable for quantitative analysis and testing.

This Chapter, then, focuses on two Research Questions:

1. *How can long-term business relationships discourage or hinder successful innovations?*
2. *How can actors' Motivations to Business Networking influence innovation performances throughout the network?*

5.2 How Can Long-Term Business Relationships Discourage or Hinder Successful Innovation?

Literature is growingly aware that the negative features of business networks have been insufficiently studied so far. Molina-Morales and Martínez-Fernández (2009) and Adler and Kwon (2002) claim that there is a lack of empirical research particularly on the negative effects of intense and trustworthy inter-firm relationships. Perks and Jeffery (2006) suggest that the apparent benefits of inter-firm collaboration in innovation processes should be critically questioned, and that the disadvantages of collaboration are under-investigated.

The few publications identifying possible negative features of business networks describe different and often serious problems. Håkansson et al. (2009) note that sometimes business networks, far from being catalyst of valuable innovation and progress, can be nests of opacity, aimed at domineeringly protect their members' interests at the expense of the society. Strong inter-firm bonding can create "inward-looking and dependence-oriented culture", according to Eklinder Frick et al. (2011). In such networks, social bonds may create organizational lock-ins and over-embeddedness, negatively influencing decision-making.

Some writings seek to identify the possible causes of negative features and poor performances of business networks. Fulconis and Pache (2008) question the researches based on the collaborative network theory, according to which it is only long-term involvement and total trust between partners that creates value. They suggest that this view is simplistic and ideological and can lead to failure; on the contrary, the authors say that a "moderate" opportunism could have a positive impact on inter-organizational networks. Another challenge to mainstream business network research comes from Staber (2001), who found that the presence of clusters of firms in the same industry increased business failure rates in the long run. In investigating the causes of poor performances of business networks, Choi (2001) seeks to find a balance between different business network theories: "Imposing too much control [to the business network] detracts from innovation and flexibility; conversely, allowing too much emergence can undermine managerial predictability and work routines. Therefore, when managing supply networks, managers must appropriately balance how much to control and how much to let emerge".

Other challenging findings on the causes of poor performances in business networks come from Knudsen (2007): "firms from their own industry tend to contribute similar knowledge, which ultimately may endanger the creation of new knowledge and therefore more radical product developments". Moreover, Knudsen found that the involvement of customers in product development had a negative impact on innovative success, probably because customers were unable to identify and to express what was actually going to be attractive to them in the future. Other explanations to network failures are offered by Pittaway et al (2004): "Where networks fail, it is due to inter-firm conflict, displacement, lack of scale, external disruption and lack of infrastructure".

Is it possible to achieve an overall, systematic view of business network success and failure as for innovation performances?

Five theoretical streams will be presented here, each of them explaining different aspects and factors explaining the possible negative influence of long-term inter-organizational relationships on innovation capabilities and innovation-related performances.

1. The Theory of the Population Ecology of Organizations (Hannan and Freeman, 1977 and 1989) asserts that organizations survive if they successfully occupy an ecological niche. The occupation of a niche usually results from successful innovations, for example from the creation of new products or novel, more efficient procedures. But once a niche is successfully occupied, the organization tends to keep its position by becoming more and more reliable and accountable, as a result of the pressure from, and the interaction with, the other actors of the business network. Thus, the same forces that drive innovation can, once success is achieved, trigger rigidity and conservatism in the long run. In fact, this theory asserts that important innovations almost always stem from the death of old organizations and from the birth of novel organizations. In other words, this theory predicts that, in networks where very reliable and accountable organizations are selected and rewarded, innovation is likely to be discouraged in the long run.

2. The Resource Dependence Theory (Pfeffer and Salancik, 1978; Hillman et al., 2009) asserts that organizations seek to enhance control on external critical resources. Either the whole network's goal is to control the external environment, like for example pressure groups do, or the stronger party seeks to control the weaker party within the network itself, like in centralized supply chains. Means to enhance control include long-term agreements and process integration (like e.g. in outsourcing contracts), new inter-organizational architectures (like e.g. in Joint Venture initiatives) and power-oriented cooperation (like in lobbying activities). This theory tends to measure network performances in terms of risk reduction and control of threats. But some writings, on the other hand, assert that too much obsession for control prevents innovation-oriented inter-organizational cooperation, which is swamped by power games, abuses, rigidities and/or over-detailed contracts (Greenhalgh, 2001). Håkansson et al. (2009) state that business

networks have a paradoxical nature: organizations tend to exert control on their networks, but control destroys the positive potentialities of the networks. In other words, this stream of studies suggests that control on external resources allows a safer and more predictable environment where investments (and then also innovation) may be encouraged; but, on the other hand, the very success of control may result in conflicts, rigidities, or privileges, that discourage innovation-oriented cooperation in the long run.

3. The Resource Based View of the firm (Combs and Ketchen, 1999) asserts that successful firms establish long-term, cooperative inter-organizational relationships when this (i) allows them to access valuable resources and (ii) allows them to develop inimitable capabilities. Scholars working with this theoretical approach often attach a great importance to inter-organizational information sharing and to the inimitable knowledge base that can be developed through long-term inter-organizational interactions. Information flows draw opportunities flows: for this reason, smooth information sharing is considered a key success factor for any innovation-oriented network.

4. The Collaborative Network Theory (Greenhalgh, 2001) asserts that in successful business networks the organizations, far from pursuing control as the key strategy for inter-firm interactions, renounce all the control-oriented strategies identified by the Resource Dependence Theory (see above, point 2 of this list), and instead rely on flexible, informal mechanisms for interaction governance, such as spontaneous resource sharing, trust, loyalty and reciprocity. This allows flexibility, co-adaptation and creative problem solving, with excellent impacts on innovation capabilities. On the other hand, criticism is arising on this view: some writings suggest that relying too much on trust and personal relationships can result in inefficient or ineffective innovation efforts, inward-looking culture (Zaheer et al., 2000) and/or in organizational fragility, since very trustful and altruist groups often attract sudden, disruptive invasions of opportunistic behaviors (Nowak, 2006).

5. The Institutional Theory (Ostrom and Walker, 2005) asserts that inter-organizational relationships are shaped by the organizations' need to be legitimated by the business community they want to be part of. As a consequence, organizations tend to follow "fashionable" management trends, to imitate the most admired organizations, to comply with the norms that consider important for their respectability, and to adapt their systems and procedures to the requests of the most important organizations they interact with. These behaviors explain organizational innovations, both the successful and the unsuccessful ones: in fact, according to this theory, managers tend to choose the innovation strategies that will give them the highest prestige in their business communities– which does not necessarily corresponds to the best possible choices.

5.3 How Can Actors' Motivations to Business Networking Influence Innovation Performances throughout the Network?

Business networks are extremely complex phenomena. An overall view of the five theories synthesized above allows the identification of at least six business network types, and related predicted innovation-related performances:

1. "Reliability / Predictability Networks": these business networks are effective in selecting accountable organizations, which respect rules and agreements and which the other partners can rely on. High predictability allows smooth resource flows, good process optimization and low transaction costs. These networks are analyzed by the Theory of the Population Ecology of Organizations mainly, that predicts that, in networks of this type, the organizations will be innovative in the first place, but, due to evolutionary drifts, will lose the capability to change in the long run.

2. "Information Exchange Networks": these business networks are effective in allowing smooth, low-cost information sharing and knowledge-oriented cooperation. The Resource Based View is particularly interested in this type of networks, whose cooperation performances are brilliantly explained by the Inverse Commons /Cornucopia of the Commons Theory, since information is often a non-rival good, that people can use and share without depleting the resource (Hess and Ostrom, 2007). This theory predicts that, provided that the benefits of using the information shared by the others are perceived as higher than the costs of participating in the network, network knowledge will boom, so benefitting the innovation-related performances.

3. "Control-Oriented Networks": in these networks, the actors' main concerns are to enhance control on the resources hold by other actors, and to limit the possibilities that their own resources and independency are controlled by other actors. If the relationships are asymmetrical, the strongest organization exerts a strict control on the other actors, for example by imposing long-term detailed contracts, or by severely punishing "naysayers", or by building joint-ventures or formal alliances. These networks may be either substantially fair to the weakest partners, or even very unfair, since in some cases control results in a sort of "inter-organizational bullying". Networks of this type are analyzed mainly by the Resource Dependence Theory, which predicts that such control-oriented network behaviors will improve firm competitive performances, at least in the short run, thorough higher efficiency and stronger risk management; influences of this type of business networking on innovation-related performances are not contemplated by the Resource Dependence Theory, whilst the Collaborative Network Theory predicts that, in this type of network, innovativeness is likely to be jeopardized in the long run.

4. "Power-Oriented Networks": in these networks, actors cooperate to enhance their control on critical resources that are external to the network, or to influence decisions that may change their context or competitive position. Lobbies, professional associations, or pressure groups are common examples. Networks of this type are analyzed by the Resource Dependence Theory, which predicts that such power-oriented network behaviors will improve the organization's profits, at least in the short run, through the network's protective control on the competitive environment; on the other hand, the organization's social performances tend to be poorer in networks of this type.

5. "Project / Change Co-Management Networks": in these networks, different organizations cooperate to co-manage specific projects or change processes. Networks of this type usually imply strong commitment and high risks, and are investigated both by the Collaborative Network Theory (which emphasizes the importance of trust, altruistic cooperation and flexibility) and by the Resource Dependence Theory (which emphasizes the importance of strong contracts and planned integration processes).

Improvements in innovation-related performances are predicted by both theories, given that some key conditions (such as, for example, trust, fairness, accountability, reliability) are met.

6. "Legitimacy Networks": organizations enter networks of this type in order to be accepted in a targeted community, and to access reliably appropriate rules, models and belief systems to conform to. Networks of this type are investigated by the Institutional Theory, that predicts that organizations will start innovation processes for possible reasons of compliance or prestige, that often do not imply innovational success: as a consequence, networks of this type usually impact on the organization's social role, much more than on its profitability or competitive performances.

In the following Paragraph, three cases of network failure will be briefly presented and discussed in the light of this framework in which 6 network types are identified, corresponding to 6 different motivations for business networking.

5.4 Three Cases of Network Failure

5.4.1 First Case: Resistances to the Introduction of ERP Systems in Supply Chains

In the year 2011, I was given the possibility to conduct a qualitative research within a leading international consulting firm focused on the implementation of ERP systems.

I conducted two semi-structured interviews with a consultant who had a several years' experience in supply chain management projects, both in Europe and South America.

For two of these projects, reports were available, that were analysed to complement the outcomes of the interviews.

"ERP systems are incredibly powerful today" said the interviewee. *"They can coordinate all the value chain activities, they can suggest how to change the production schedules to match the emerging demands of the market, they can automatically spot and correct errors in sales forecasting, they can make warehouse costs incredibly lower... all this notwithstanding, people puts up exasperating resistance against ERP projects, especially partnering firms".*

The interviewee was strongly convinced that ERP systems could provide substantial help in addressing the complex challenges of supply chain management, but she thought that there are structural reasons for which the industrial networks involved in ERP-based innovation projects put up resistance against this type of innovation.

"ERP systems are usually implemented within the strongest company of the value chain, or within the company occupying a key position within the chain. The other partners feel that they may be forced to introduce costly changes in their routines without controlling what they are paying for, and that they risk to find themselves chained up to the strongest partner" said the interviewee. In other words, partners who are expected to accept and to

comply with the new ERP system controlled by the major company are afraid that such a radical innovation would make the "quit-the-network option" impossible to them: this would make them weaker and weaker in negotiations with the bigwig, i.e. the ERP owner.

"In one of the supply chain projects I coordinated, people of one organization (which was the purchaser in the value chain) were expected to entry data that were to be used by the partnering organization (the supplier), owner of the ERP system, in order to schedule their production processes. Since some data entry errors occurred, the supplier claimed that people in the purchasing organization who had failed in effectively entering data were punished or substituted. This was interpreted as a sort of organizational invasion on the part of the purchasing organization, and resulted in conflicts that jeopardized the project".

What comes out is that when an innovation launched at the network level implies internal organizational changes and/or high investments, the possible range of relational difficulties gets much wider. *"There must be a proportion between the investment an organization makes for a certain innovation, and the degree of control that can exert on it. If the proposed innovation implies that one of the partnering organizations becomes weaker in possible future conflicts, this organization will probably put up resistance unless compensations are offered".*

This case is an interesting example of Project / Change Co-Management Network. As detailed in Paragraph 5.3, networks of this type are disputed between the Resource Dependence Theory and the Collaborative Network Theory.

Here, the two partners proved incapable of solving a conflict, both through contracts (as suggested by the Resource Dependence Theory), and through cooperation and mutual understanding (as suggested by the Collaborative Network Theory): this caused innovation failure. This case confirms the predicted importance of factors that enhance the sustainability of long-term cooperation within these networks; these factors have been operationalized in Chapter 3, through the variable Business Network Strength.

5.4.2 Second Case: the Breakdown of the Textile District of Prato, Italy

Edoardo Nesi, a former entrepreneur in the textile industry, published his autobiographic novel "Storia della mia gente" (that means "Story/History of my People") in 2010. In this novel, he describes the flowering and the collapse of the textile district of Prato, near Florence, in Italy. A translation of some particularly meaningful passages of the book follows.

The Golden Age

"Imagine a product that has not needed any change for thirty years. Imagine a company that makes that product only, and whose sole problem is being incapable of manufacturing enough of it to meet the demand of a market, which is so huge and viable that the impact of competitors is negligible. Imagine you can set your clock on the punctuality of your customer's payments, ten days after invoice date, no claims, no pretexts, no bankruptcies; the cheques arrive every morning, in light blue envelops. Wipe out any costs of research and

development, of trade fairs, of advertising, of stylistic advisory. Cancel the concept of leftover stock. Laugh fit to burst of the idea of hiring a manager to do the job you can perfectly do on you own.

And now, imagine a whole town founded on textile industry, starred with dozens and dozens of firms like yours, all continuously growing and all interconnected in a system foolishly fragmented but incredibly effective, made up of hundreds of often family-owned microbusinesses, each of them focused on a specific in-between phase of manufacturing, each of them with its name, its pride, its profits". (p. 26).

The Impact of Globalization

"Those were the days when I was still angry, those days around the new millennium, when the revenues of our business continued to decrease, year after year, month after month, and I used to return home full of rage for the auctions to which our customers forced us by that time, attaching no more importance to the quality of our fabric, to the reliability of our services, to the punctuality of our deliveries, to the name of our business and to its history. It seemed as if the customers had become all deaf, all of them, even the Germans.

It was only the price that mattered, and in matter of price we always lost, because there was always someone more desperate than us [...] The auction [...] went on, stupid and wicked, and the moment arrived in which the desperate entrepreneur had to tidy up his hair and to get into his Mercedes ML or his Audi [...] and go to strangle artisans even more desperate than him, those that would have to spin and to weave to offer an even lower price to the customer, in a pernicious spiral that showed the dirty side of the idea of free market" (p. 60).

"We should have protected ourselves against globalization!"

"This is our story.

The story of millions of people, betrayed also and above all by our politicians, who [...] had silently put their signatures at the bottom of hundreds of treaties that would scalp the Italian manufacturing industry. [...] It is February 28th, 2009, and I have taken to the streets for the first time in my life [...] There are many entrepreneurs among the demonstrators [...] warmed by the novel idea that Prato is not just a galaxy of small and fervidly individualistic businesses, as it has always been, each of them facing the world and the sharp destiny, but a real economic community, capable to gather and to speak with one voice, even if for one single day [...] But it is difficult to identify a recipient of the protest, a culprit, a villain, other than the status quo of the world, and this concrete, calm protesting of thousands of people living in the same town but divided by everything, against the very substance of things, seems sublime to me, and unprecedented, and proudly useless" (pp. 138 – 150).

As the reader can see, Edoardo Nesi's family business was embedded both in a vertical network of local suppliers and international customers, and in a horizontal network of partners and competitors in the territorial cluster of Prato: but neither network was capable to help when the axe of globalization stroke. After decades of perfect fairness, the firm's customers suddenly transformed themselves in pitiless opponents. The firms of the Prato district, after decades of "fervid individualism" in which networking just meant to fully respect the gentlemen's agreements between firms, where completely incapable to cooperate to pursue a solution together. They gathered only when the battle was already lost, and for one day's street protest only. Nesi just complained that Prato entrepreneurs had been unable to lobby the politicians against the treaties that had opened the markets to Chinese competitors. In the whole novel, not a single attempt of co-innovation, not a single effort of sharing resources to find a common solution, not a single proactive meeting with another actor, either of the vertical chain or of the horizontal cluster, is mentioned.

This case depicts a typical Reliability / Predictability Network (see Paragraph 5.3) and corroborates the predictions of theory of Population Ecology of Organizations, according to which long-term, repetitively fair interactions are a double-blade weapon: they encourage specialization and they reduce transaction costs even dramatically, but they result in a sort of "immune weakness" against possible sudden changes in the partners' requests and attitudes, in the future. Even more importantly, networks of this type are very conservative since the resources and capabilities to innovate are perceived as useless, and tend to atrophy. When the organization's clients suddenly shifted from behaving as Reliability / Predictability Network partners to Control Oriented partners (see Paragraph 5.3), the organization was jeopardized, and when eventually all the firms of the districts decided to attempt to create a Power-Oriented Network (see Paragraph 5.3) to protect themselves, it was too late.

5.4.3 Third Case: the Destruction of Landscape Resources in Alpine Tourism Destinations

This study developed through several years, because I have been participating in some activities of an important international ONG aimed at the sustainable development of the Alpine region since 2002-2003. I have been regularly receiving the ONG's newsletter, I have been participating in several conferences organized by the Italian branch of the ONG, based in Turin, and I have been regularly reading the related reports and press releases. Moreover, I repeatedly conducted informal interviews with national and international speakers invited at the conferences. I could then see how ideas on factors of success and failure of the Alpine region as a system of tourism destinations were being discussed in the last decade, and I could witness the evolution of the Alpine destination performances in a medium-term period of time.

In these years, I could collect several elements that led me to reflect upon the possible negative role of local tourism business networks. Hotels, restaurants, tourist attractions and other facilities of several Alpine destination, in fact, have formed powerful formal and informal networks, which have sometimes been strongly influencing the Public Administration bodies and the political leaders about urban planning and infrastructure strategies for decades.

"When a hotel or a restaurant planned to over-exploit the landscape, for example by high-impact building or polluting, the other actors often did not protest, because they did not want to be hampered, in turn, in their future initiatives" complained a manager of a local Public Administration body during a conference. *"They thought that tourists would continue to come by, no matter how deeply the landscape was compromised"*.

Those networks of tourism stakeholders, in other words, have been acting as lobbies and were substantially aimed at allowing the utmost exploitation of the landscape resources on the part of network members. *"If a tourism entrepreneur sees that the neighbours opportunistically exploit the landscape for their own immediate interest, and, say, the hotel association X facilitates and protects such opportunistic behaviours, the entrepreneur will hasten to do the same, in the fear that someone else exploits the opportunities before"* said a manager of the ONG. The individual advantage from exploiting common resources is often perceived as being greater than the potential long-term shared losses that result from the deterioration of such resources. Thus, there is little motivation for individual actors (whether governments, elected officials, or individual operators), to invest or engage in concrete, costly projects for a more sustainable tourism. The environmental resources are then destroyed, sometimes in a virtually irreversible way, and this results, sooner or later, in the decline of the tourism destination (Buhalis, 2000).

This case depicts a Power-Oriented Network (see Paragraph 5.3) and confirms the prediction of higher profits in the short run, and worse social impact in the medium and long run, of organizations choosing this type of networking strategy, where innovation is not likely to play an important role.

5.5 Measuring the Motivations for Business Networking

As I sought to depict in Paragraph 5.3, and to corroborate in Paragraph 5.4, there are different types of networks, that attract (or are built by) organizations with different goals and interests, and that tend to result in different innovation-based performances.

For example, if a network is effective in providing participants with valuable information, it will attract organizations interested in enhancing their knowledge base; this type of network is likely to perform as predicted by the resource Based View, and by the Cornucopia of the Commons Theory (see Paragraph 5.3).

Thus, the extent to which a certain business network is perceived as effective in one of the 6 aspects identified in Paragraph 5.3 can be considered, I suggest, also a good proxy of the organization's motivations for business networking, and a good predictor of short-term and long-term innovation-based performances.

For this reason, I decided to build 6 Scales, each of them corresponding to one of the 6 network types identified in Paragraph 5.3.

The Scales were built on the basis of literature analysis and theoretical reflections, on the one side, and of the qualitative research described in Paragraph 5.4, on the other side. They were then discussed and fine-tuned with the five experts involved in this research (see Chapter 1). The results follow.

5.5.1 Importance of Reliability/Predictability as a Motivation for Business Networking (NM-R)

5 Items Scale

Do you agree with the following statements? Please consider your organization's situation in the last three years. (1=strongly disagree, 5=strongly agree, 3=neutral).

NM-R-1
In our business context, it is often better to deal with trustful and reliable counterparts, even at the cost of renouncing attractive alternative proposals or possibilities.
NM-R-2
A great deal of our organization's success stems from our policy of excluding potentially unreliable subjects from our business relationships, even if they could be potentially interesting.
NM-R-3
Our organization's success is strongly rooted in our reputation of being trustful, fair and rule-abiding.
NM-R-4 (reverse item)
In our milieu, business relationships tend to be very numerous, extemporary and impersonal: resources are invested in finding possible substitutes (e.g. new customers, new suppliers, etc.) more than in the existing relationships.
NM-R-5
In our business context, customers are usually very loyal once good product quality is assured.

5.5.2 Importance of Information Exchange as a Motivation for Business Networking (NM-I)

6 Items Scale

Do you agree with the following statements? Please consider your organization's situation in the last three years. (1=strongly disagree, 5=strongly agree, 3=neutral).

NM-I-1
In our milieu, a good network of relationships allows access to business-relevant information and advice, which otherwise would be hardly available.

NM-I-2

Most of the information allowing true competitive advantage for our organization were accessed through our organization's network of relationships.

NM-I-3

Our organization invests to provide managers and employees with occasions (e.g. participation in trade fairs, conferences, associations, clubs) to share valuable information with established or potential business partners.

NM-I-4

Our managers are encouraged to leverage our business network to collect valuable information and to make it available for the whole organization.

NM-I-5

Our organization supports associations or organizations specifically aimed at business-related sharing of information and knowledge.

NM-I-6

Thanks to our long-term business relationships, our organization gained in-depth knowledge of many important actual and prospect business partners, of their behaviors, needs, cultures, structures, markets, etc.

5.5.3 Importance of Control of Network Resources as a Motivation for Business Networking (NM-C)

4 Items Scale

Do you agree with the following statements? Please consider your organization's situation ***in the last three years****. (1=strongly disagree, 5=strongly agree, 3=neutral).*

NM-C-1

In our milieu, the weakest partners in B2B networks (e.g., small suppliers) are often over-exploited and swamped by the strongest organizations (e.g. key clients, or leading dealers/intermediaries).

NM-C-2

Our organization is, or will probably be, strongly bonded to another company through acquisition, board interlocks, joint-venture or long-term formal alliance.

NM-C-3

In our milieu, important resources must be dedicated to accurately define and legally secure the organization's relationships with the other parties, for example through long-term contracts, detailed service level agreements, etc.

NM-C-4 (reverse item)

In our milieu, contracts are sometimes considered as just sheets of paper, or as annoying factors of rigidity: trust and good personal relationships are often more effective.

5.5.4 Importance of Power on External Resources and Decisions as a Motivation for Business Networking (NM-P)

5 Items Scale

Do you agree with the following statements? Please consider your organization's situation in the last three years. (1=strongly disagree, 5=strongly agree, 3=neutral).

NM-P-1 (reverse item)
In our organization's milieu, a well-structured, competitive and innovative organization usually succeeds in business, even without special protections (such as political support or friends in high places).

NM-P-2
Political decisions may impact our sector heavily: joining forces also with competitors, to make pressures against damaging decisions, is a good investment.

NM-P-3
In our sector, the main role of professional / industrial associations should be protecting the members against system threats such as excessive bureaucracy, excessive taxes, or excessive competition.

NM-P-4
In our organization's milieu, a potential business partner (e.g. a supplier, a distributor, a professional firm) accredited as having friends in the high places is often preferred to a more competent and reliable, but not so well-connected subject.

NM-P-5 (reverse item)
In our organization's milieu, the procedures for getting a positive response from the Public Administration (e.g. authorizations, public funding, subsidies, verdicts) are efficient, transparent and equal for all.

5.5.5 Importance of Project / Change Co-Management as a Motivation for Business Networking (NM-M)

5 Items Scale

Do you agree with the following statements? Please consider your organization's situation in the last three years. (1=strongly disagree, 5=strongly agree, 3=neutral).

NM-M-1
In our industry, many of the most successful organizations have effectively integrated their processes with their main customers and/or suppliers and/or partners, for example by sharing databases, or by co-automatizing order processing.

NM- M -2

A great deal of our organization's success stems from innovation projects we carried on in cooperation with other subjects, such as e.g. suppliers, customers, research institutions, Public Administration bodies, banks, etc.

NM- M -3

Our organization successfully relies on third parties for critical activities, such as, for example, IT innovation, market analyses or internationalization projects.

NM- M -4 (reverse item)

In our milieu, the organizations rarely cooperate on specific projects, because the organizations often don't like the idea of sharing confidential processes and information.

NM-M-5 (reverse item)

In our organization's milieu, the organizations rarely cooperate on specific projects, because people often don't like the idea of the reciprocal dependence that process and/or project integration may result in.

5.5.6 Importance of Legitimacy as a Motivation for Business Networking (NM-L)

4 Items Scale

Do you agree with the following statements? Please consider your organization's situation ***in the last three years****. (1=strongly disagree, 5=strongly agree, 3=neutral).*

NM-L-1

In our industry, becoming a supplier or a partner of a leading company can boost an organization's reputation even dramatically.

NM- L -2

In our industry, there are some recognized best practices for innovation, that most organizations seek to imitate.

NM- L -3

If an organization wants to be legitimated as reliable in our sector, it must comply with a complex system of norms and procedures.

NM- L -4 (reverse item)

In our milieu, smaller or younger organizations tend to adapt to a leading organization's culture, practices and systems, so that their interactions with the leader are smoother and stronger.

References

Adler, P., & Kwon, S-W. (2002). Social capital – prospect for a new concept. Academy of management review, 27 (1) 17-40.

Bànàthy B.H. (1996), Designing Social Systems in a Changing World. Plenum, NY

Björk, J. & Magnusson, M. (2009). Where Do Good Innovation Ideas Come From? Exploring the Influence of Network Connectivity on Innovation Idea Quality. Journal of Product Innovation Management, 26(6): 662–670.

Brandon, R. and Burian R.M. eds., (1984). Genes, Organisms, Population: Controversies Over the Units of Selection. Cambridge MA: MIT Press

Brandt H., Ohtsuki H, Iwasa Y., and Sigmund K. (2006). A Survey on Indirect Reciprocity. IIASA Interim Report, IR-06-065. IIASA Studies in Adaptive Dynamics, n.123.

Buhalis, D., (2000), Marketing the competitive destination of the future, Tourism Management, Vol.21(1), pp.97-116.

Choi, T.I. (2001). Supply networks and complex adaptive systems: control versus emergence. Journal of Operations Management; May2001, Vol. 19 Issue 3, p351-366.

Combs J.G., Ketchen D. (1999). Explaining Interfirm Cooperation and Performance. Toward a reconciliation of predictions from the resource-based view and organizational economics. Strategic Management Journal, 20:867-888.

Eisenhardt K.M. and Bird Schoonhoven C. (1996). Resource-Based View of Strategic Alliance Formation: Strategic and Social Effects in Entrepreneurial Firms. Organization Science, Vol. 7, No. 2 (Mar. - Apr., 1996), pp. 136-150

Eklinder Frick J., Eriksson L.T., and Hallén L. (2011). Bridging and bonding forms of social capital in a regional strategic network. Industrial Marketing Management, Vol. 40, Issue 6, August 2011, pp. 994 – 1003.

Fromm, J. (2004), The Emergence of Complexity. Kassel University Press

Fulconis, F., Pache, G. (2008). Le management strategique des reseaux inter-organisationnels a l'epreuve des comportements opportunistes: Elaboration d'un cadre d'analyse. La Revue des Sciences de Gestion, March-April 2008, v. 43, iss. 230, pp. 35-43.

Gould, J. (1980). The panda's thumb. New York: Norton.

Gould, S.J. (2002). The Structure of Evolutionary Theory. Harvard University Press.

Greenhalgh, L. (2001). Managing strategic relationships: The key to business success. Free Press.

Gulati, R. (2007). Managing network resources. Alliances, Affiliations and Other Relational Assets. Oxford University Press.

Håkansson, H., Ford, D., Gadde, L-E., Snehota I, and Waluszewski A.(2009), Business in Networks. Chichester: Wiley.

Hart C. (2000). Doing a literature review. Sage Publications: London

Hindmoor A. (1998). The Importance of Being Trusted: Transaction Costs and Policy Network Theory. Public Administration, Vol. 76, Issue 1, pp. 25-43.

Knudsen, M.P. (2007). The Relative Importance of Interfirm Relationships and Knowledge   Transfer for New Product Development Success. Journal of Product Innovation Management; Mar2007, Vol. 24 Issue 2, p117-138.

Hannan, M.T. and J. Freeman (1977) "The population ecology of organizations." American Journal of Sociology 82 (5): 929-964.

Hannan, M.T. and J. Freeman (1989) Organizational Ecology. Cambridge, MA: Harvard University Press.

Hess C., Ostrom E. (eds.) (2007). Understanding Knowledge as a Commons. MIT Press.

Hillman, A. J., Withers, M. C. and B. J. Collins (2009). "Resource dependence theory: A review." Journal of Management 35: 1404-1427.

Lieberman E., Hauert C., and Nowak M.A.(2005). Evolutionary dynamics on graphs. Nature 433: 312-316

Molina-Morales, F., Martínez-Fernández, T. (2009). Too much love in the neighborhood can hurt: How an excess of intensity and trust in relationships may produce negative effects on firms, Strategic Management Journal, 30, 1013-1023.

Nesi E. (2010). Storia della mia gente. Milano: Bompiani.

Nowak M.A. (2006), Five Rules for the Evolution of Cooperation. Science, 314, 8: 1560-1563.

Ohtsuki H., C HauertC., Lieberman E., and Nowak M.A. (2006). A simple rule for the evolution of cooperation on graphs and social networks. Nature 441: 502-505

Ostrom E. and Walker J. (2005). Trust and reciprocity: Interdisciplinary Lessons for Experimental Research. Russel-Sage Foundation.

Perks, H., & Jeffery, R. (2006). Global network configuration for innovation: a study of international fibre innovation. R&D Management , pp.67-83.

Pfeffer, J. and G. R. Salancik (1978). The External Control of Organizations: A Resource Dependence Perspective. New York, NY, Harper and Row.

Pittaway L., Robertson M., Munir K., Denyer D., Neely A. (2004). Networking and innovation: a systematic review of the evidence. International Journal of Management Reviews, Volume 5, Issue 3-4, pages 137–168, September 2004.

Ricciardi F., Cantù C. (2011). The Role of Altruism in Inter-firm Relationships: Long-term Value Creation in Business Networks. The paper was published at the 27th IMP-conference in Glasgow, Scotland in 2011. Available online at: http://www.impgroup.org/uploads/papers/7754.pdf .

Sigmund K. (1998). Complex Adaptive Systems and the Evolution of Reciprocation. IIASA Interim Report, IR-98-100. IIASA Studies in Adaptive Dynamics, n.32.

Staber, U. (2001). Spatial Proximity and Firm Survival in a Declining Industrial District: The Case of Knitwear Firms in Baden-Württemberg. Regional Studies; Jun2001, Vol. 35 Issue 4, p329-341

Trivers R.L. (1971). The evolution of Reciprocal Altruism. The Quarterly Review of Biology, 46 (1): 35-57.

Tushman, M.L., and Nadler D. (1986) Organizing for innovation: towards successful translational research. California Management Review, Vol. 27, Issue 10, 558-61.

Von Bertalanffy L. (1976) General System theory: Foundations, Development, Applications. George Braziller.

Walsham G. (2006). Doing interpretive research. European Journal of Information Systems (2006) 15, 320–330.

Westerlund, M. & Rajala, R. (2010). Learning and innovation in inter-organizational network collaboration. Journal of Business & Industrial Marketing, 25(6): 435-442.

Yin R. (1984), Case study research: Design and methods (1st ed.). Sage.

Zaheer, A., Gulati, R., & Nohria, N. (2000). Strategic networks. Strategic Management Journal, 21(3), 203.

6. Innovational Effectiveness of Different Network Types: Theory Building and Testable Models

Abstract This final Chapter utilizes the findings of the previous five Chapters to answer the Research Question of this book: how does business networking influence the organization's innovation processes and performances? The theory-building outcomes of the previous Chapters are used to build six models linking the six main Motivations for Business Networking identified in Chapter 5 to the main innovation-related capabilities and performances identified in Chapters 2 and 4, through the important moderating role of network long-term fairness and cooperation, analysed in Chapter 3. These six testable Models, describing six different network types, are associated to a synthetic theoretical explanation of the hypothesized cause-effect relationships, and include all the 12 constructs developed and operationalized in this book.

6.1 Introduction

In Chapter 5, six different types of business networks were identified, each of them corresponding to a specific expected network effectiveness, and then to a specific motivation for business networking. These 6 types of business networks are synthesized in Table 6.1.

Table 6.1. The Six Different Types of Business Networks

No.	Type	Description	Theories Involved
1	Reliability/Predictability Networks	Organizations concentrate on those reliable relationship that allow process predictability.	Population Ecology of Organizations.
2	Information Exchange Networks	Organizations concentrate on those relationships that allow access to valuable information and knowledge.	Resource Based View. Game/Systems Theories (Inverse Commons Th.)
3	Control Oriented Networks	Organizations manage relationships in order to enhance control on critical resources owned or controlled by other network actors.	Resource Dependence Theory. Collaborative Network Theory.
4	Power Oriented Networks	Organizations manage relationships in order to enhance control on external resources and/or political decisions.	Resource Dependence Theory. Transparency and Social Capital studies.
5	Project/Change Co-Management Networks	Organizations share innovation project with selected network partners, and cooperate in order to find solutions and implement changes.	Collaborative Network Theory. Resource Dependence Theory.
6	Legitimacy Networks	Organizations concentrate on those relationships that allow to be accepted in a targeted community.	Institutional Theory.

Literature and qualitative research confirmed that these six motivations for business networking attract and create different behaviors and impact innovation processes even very differently.

Of course, cases of overlapping between these six network types are possible in the real world: the same business network may have two-fold effectiveness, for example as for Information Exchange (see network type 2 in Figure 6.1) and Power (network type 4), like in many typical lobbies.

Moreover, each organization can be part of different networks at the same time: for example, an organization may interact with a restricted network for Project Co-Management (network type 5), and with another, larger network for Legitimacy (network type 6).

In these cases, the impacts of the different motivations for business networking sum up, sometimes reinforcing, sometimes weakening each other; but in order to understand these complex dynamics, it is necessary to understand how each of the six motivations for business networking, taken in isolation, influences innovation processes and performances.

For this reason, I developed 6 models (see Figures from 6.1 to 6.6), each of them corresponding to one of the 6 network types in its "pure" form.

These models, which will be discussed in the following Paragraphs, link through cause-effect relationships the 12 constructs developed and operationalized in the previous Chapters of this book. These 12 constructs are synthesized in Table 6.2, and the related Scales are presented in Tables from 6.3 to 6.14.

Table 6.2. The Twelve Constructs Involved in the Modular Model Proposed

Code	Name	Developed in	Scale: see
SS	Strategic Support to Innovation	Chapter 4	Table 6.3
IC	Organization's Innovation Capabilities	Chapter 4	Table 6.4
IS	Innovational Success	Chapter 2	Table 6.5
CE	Competitive Excellence	Chapter 2	Table 6.6
SE	Social Excellence	Chapter 2	Table 6.7
NS	Network Strength	Chapter 3	Table 6.8
NM-R	Importance of Reliability/Predictability as a Motivation for Business Networking	Chapter 5	Table 6.9
NM-I	Importance of Information Exchange as a Motivation for Business Networking	Chapter 5	Table 6.10
NM-C	Importance of Control of Network Resources as a Motivation for Business Networking	Chapter 5	Table 6.11
NM-P	Importance of Power on External Resources and Decisions as a Motivation for Business Networking	Chapter 5	Table 6.12
NM-M	Importance of Project / Change Co-Management as a Motivation for Business Networking	Chapter 5	Table 6.13
NM-L	Importance of Legitimacy as a Motivation for Business Networking	Chapter 5	Table 6.14

All the models have a common structure, and some of the hypothesized cause-effect relationships are the same in all 6 cases. In all Figures, from 6.1. to 6.6, in fact, it is hypothesized (consistently with the findings of Chapters 2 and 4) that the Strategic Support to Innovation (SS) positively influences the Organization's Innovation Capabilities (IC), which positively influences Innovational Success (IS), which positively influences Competitive Excellence (CE); moreover, also the variable Social Excellence (SE) is included, to assess the ethical results of the organization's innovations (or conservatism).

These five variables (SS, IC, IS, CE, SE, see Table 6.2 and Tables from 6.3 to 6.7) are identified with white boxes in Figures from 6.1 to 6.6. They were designed to identify and measure the five core innovation-related phenomena that were suspected to be influenced by business networking processes.

Each of the six models drawn in this Chapter hypothesizes the possible different influences of Network Strength (see Chapter 3) and of the specific Motivation for Networking under examination (see Table 6.1 and Chapter 5) on the five innovation-related variables identified. Network Strength and Motivations for Networking (NS and MN, see Table 6.2 and Tables from 6.8 to 6.14), whose influences change in each of the 6 models, are identified with the grey boxes in Figures from 6.1. to 6.6.

In other words, in this Chapter I propose a general model on how business networking influences innovation performances. This model is made up of 6 sub-models, each of them representing the specific structure of cause-effect relationships in one of the 6 network types described in Table 6.1. I propose that this modular model is useful to explain the even very different performances of business networks, in many cases in which the traditional variables, such as Trust or Resource Flows, are ranking similarly high.

6.2 Reliability / Predictability Networks

These business networks are effective in selecting accountable organizations, which respect rules and agreements and which the other partners can rely on. High predictability allows smooth resource flows, good process optimization and low transaction costs. Organizations need to develop innovation capabilities in the first place, to conquer and to keep their ecological niche; but the Theory of the Population Ecology of Organizations predicts that, in networks of this type, successful organizations are incentivized to replicate previously successful strategies and to standardize and/or specialize more and more (see Chapter 5). As a consequence, in networks of this type the organizations tend to become more and more rigid and incapable to change in the long run. I discussed this with the five experts involved in this research (see Chapter 1), and we agreed that, considering the specific Items of the Scales characterizing the different innovation-related processes and performances, the variables affected in this type of networks are likely to be Innovation Capabilities (IC) and Competitive Excellence (CE) (Figure 6.1). Consistently with the case described in Paragraph 5.4.2 and with the Theory of the Population Ecology of Organizations, I hypothesize that Reliability / Predictability as a Motivation for Business Networking tends to positively influence Innovation Capabilities and Competitive Excellence in the short run, but to negatively influence them in the long run. Moreover, I hypothesize that Business Network Strength (NS), being a strong stabilizing factor, tends to magnify both the positive influence of this Motivation for Business Networking in the short run, both, regrettably, also its negative influence in the long run.

6.3 Information Exchange Networks

These business networks are effective in allowing smooth, low-cost information sharing and knowledge-oriented cooperation. The Inverse Commons /Cornucopia of the Commons Theory (see Chapter 5) predicts that, provided that the benefits of using the information shared by the others are perceived as higher than the costs of participating in the network, network knowledge will boom, so benefitting the innovation-related performances. I discussed this with the five experts involved in this research (see Chapter 1), and we agreed that, considering the specific Items of the Scales characterizing the different innovation-related processes and performances, the variable affected in this type of networks is likely to be Innovational Success (IS) (Figure 6.2). This hypothesis was formulated on the basis of literature only, since it is highly consistent in predicting positive performances of this type of network (see Chapter 5); the three worst cases analyzed in Paragraph 5.4 did not include any Information Exchange Network.

6.4 Control Oriented Networks

In these networks, the actors' main concerns are to enhance control on the resources hold by other actors, and to limit the possibilities that their own resources and independency are controlled by other actors.

If the relationships are asymmetrical, the strongest organization exerts a strong control on the other actors, for example by imposing long-term detailed contracts, or by inter-organizational bullying. Fairer or more symmetrical relationships of this type typically result, for example, in joint-ventures or formal alliances. The Resource Dependence Theory predicts that such control-oriented network behaviors will improve firm competitive performances, at least in the short run, thorough higher efficiency and stronger risk management; but the Collaborative Network Theory predicts that this strategy is detrimental in the long run. I discussed this with the five experts involved in this research (see Chapter 1), and we agreed that, considering the specific Items of the Scales characterizing the different innovation-related processes and performances, the variable affected in this type of networks is likely to be Competitive Excellence (CE), whilst Innovation Capabilities and Innovational Excellence are not (directly) impacted. (Figure 6.3). Consistently with extant literature, I hypothesized that Control as a Motivation for Business Networking tends to positively influence Competitive Excellence in the short run, but to negatively influence this variable in the long run. Moreover, also in this network type I hypothesize that Business Network Strength (NS), being a strong stabilizing factor, tends to magnify both the positive influence of this Motivation for Business Networking in the short run, both, regrettably, also its negative influence in the long run. This is consistent with the findings of the case study described in Paragraph 5.4.1.

6.5 Power Oriented Networks

In these networks, actors cooperate to enhance their control on critical resources that are external to the network, or to influence decisions that may change their context or competitive position. Lobbies, professional associations, or pressure groups are common examples. Networks of this type are analyzed by the Resource Dependence Theory, which predicts that such power-oriented network behaviors will improve the organization's profits, at least in the short run, through the network's protective control on the competitive environment; on the other hand, the organization's social performances tend to be poorer in networks of this type. Moreover, the Collaborative Network Theory predicts that this strategy is detrimental in the long run also for profits and competitiveness. I discussed this with the five experts involved in this research (see Chapter 1), and we agreed that, considering the specific Items of the Scales characterizing the different innovation-related processes and performances, the variable affected in this type of networks is likely to be Competitive Excellence (CE) and Social Excellence (SE) (Figure 6.4) whilst Innovation Capabilities and Innovational Excellence are not (directly) impacted.

Consistently with extant literature, I hypothesized that also Power as a Motivation for Business Networking tends to positively influence Competitive Excellence in the short run, but to negatively influence this variable in the long run. On the other side, I hypothesize, consistently with studies on lobbying, network opacity and mob ties (see Chapters 2 and 5), that Power as a Motivation for Business Networking tends to negatively influence Social Excellence. Moreover, also in this network type I hypothesize that Business Network Strength (NS), being a strong stabilizing factor, tends to magnify both the positive influence of this Motivation for Business Networking in the short run, both, regrettably, also its negative influence in the long run. This is consistent with the findings of the case study described in Paragraph 5.4.3.

6.6 Project / Change Co-Management Networks

In these networks, different organizations cooperate to co-manage specific projects or change processes. Networks of this type usually imply strong commitment and high risks, and are usually built on the basis of strong strategic commitment. These business environments are investigated both by the Collaborative Network Theory (which emphasizes the importance of trust, altruistic cooperation and flexibility) and by the Resource Dependence Theory (which emphasizes the importance of strong contracts and planned integration processes). Improvements in innovation-related performances are predicted by both theories.

I discussed this with the five experts involved in this research (see Chapter 1), and we agreed that, considering the specific Items of the Scales characterizing the different innovation-related processes and performances, the variables affected in this type of networks are likely to be Strategic Support to Innovation (SS), Innovation Capabilities (IC) and Innovational Success (IS) (Figure 6.5). Consistently with literature and with the success cases told by my interviewee in the case described in Paragraph 5.4.1, I hypothesize that the Strategic Support to Innovation (SS) positively influences Project / Change Co- Management as a Motivation for Business Networking, which, in turn, positively impacts Innovational Success (IC). The latter positive relationship is positively moderated by Innovation Capabilities (IC). Moreover, I hypothesize that in this type of network Network Strength (NS) positively moderates the positive relationship between Strategic Support to Innovation (SS) and Innovation Capabilities.

But, as showed in the case described in Paragraph 5.4.1, this virtuous circle is quite fragile. If inter-organizational conflicts arise for controlling resources, for example because of opportunism or poor contracts, the whole project can be jeopardized. The role of Network Strength as a moderator of the relationship between Strategic Support to Innovation (SS) and Project / Change Co- Management as a Motivation for Business Networking is probably very strong.

6.7 Legitimacy Networks

Organizations enter networks of this type in order to be accepted in a targeted community, and to access reliably appropriate rules, models and belief systems to conform to. Networks of this type are investigated by the Institutional Theory, that predicts that organizations will start innovation processes for possible reasons of compliance or prestige, that often do not imply innovational success (see Chapter 5): as a consequence, networks of this type usually impact the organization's social role, much more than its profitability or competitive performances.

I discussed this with the five experts involved in this research (see Chapter 1), and we agreed that, considering the specific Items of the Scales characterizing the different innovation-related processes and performances, the variable affected in this type of networks is likely to be Social Excellence (SE) (Figure 6.6).

Consistently with extant literature, I hypothesized that Legitimacy as a Motivation for Business Networking tends to positively influence Social Excellence, especially if network transparency and accountability are high. Moreover, I hypothesize that Business Network Strength (NS), being a strong stabilizing factor, positively moderates this positive relationship.

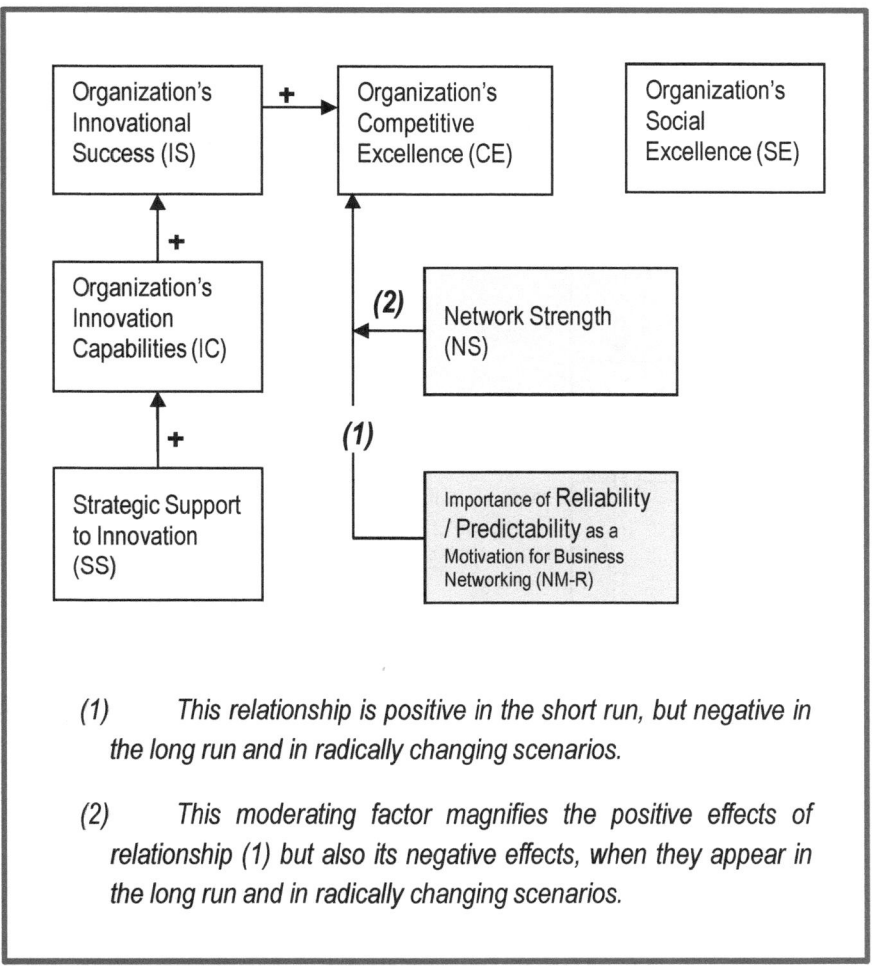

(1)   This relationship is positive in the short run, but negative in the long run and in radically changing scenarios.

(2)   This moderating factor magnifies the positive effects of relationship (1) but also its negative effects, when they appear in the long run and in radically changing scenarios.

Fig. 6.1 Influence of Business Networking (measured through Network Reliability / Predictability as a Motivation and through Network Strength) on innovation-related processes and performances

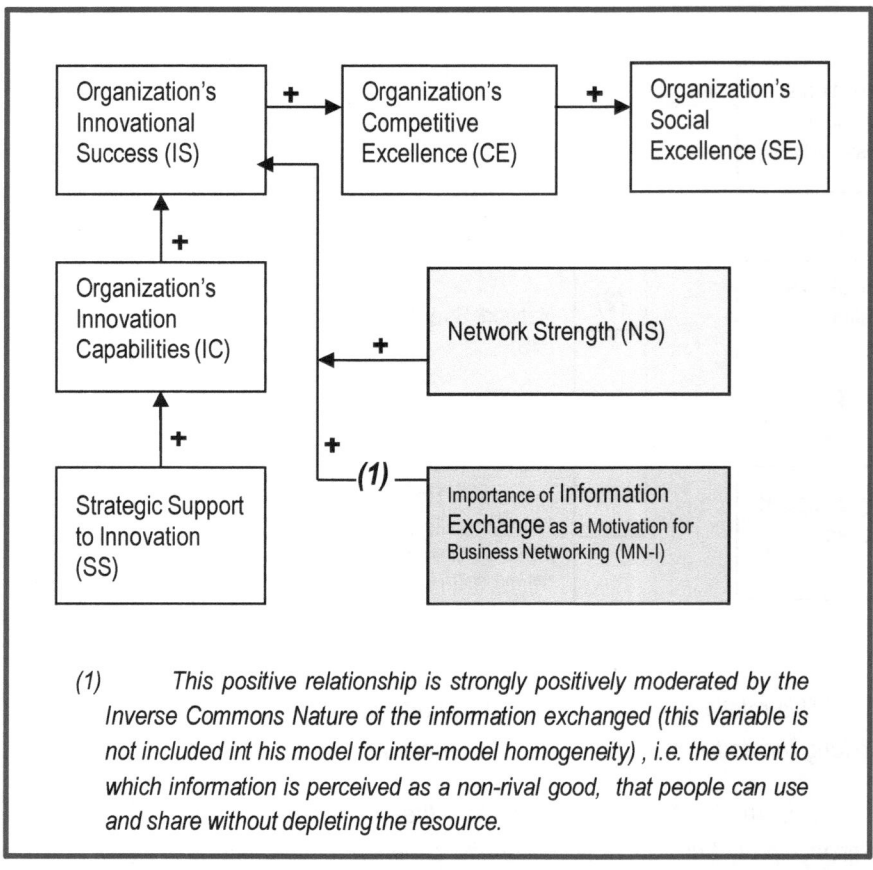

Fig. 6.2 **Influence of Business Networking (measured through Information Exchange as a Motivation and through Network Strength) on innovation-related processes and performances**

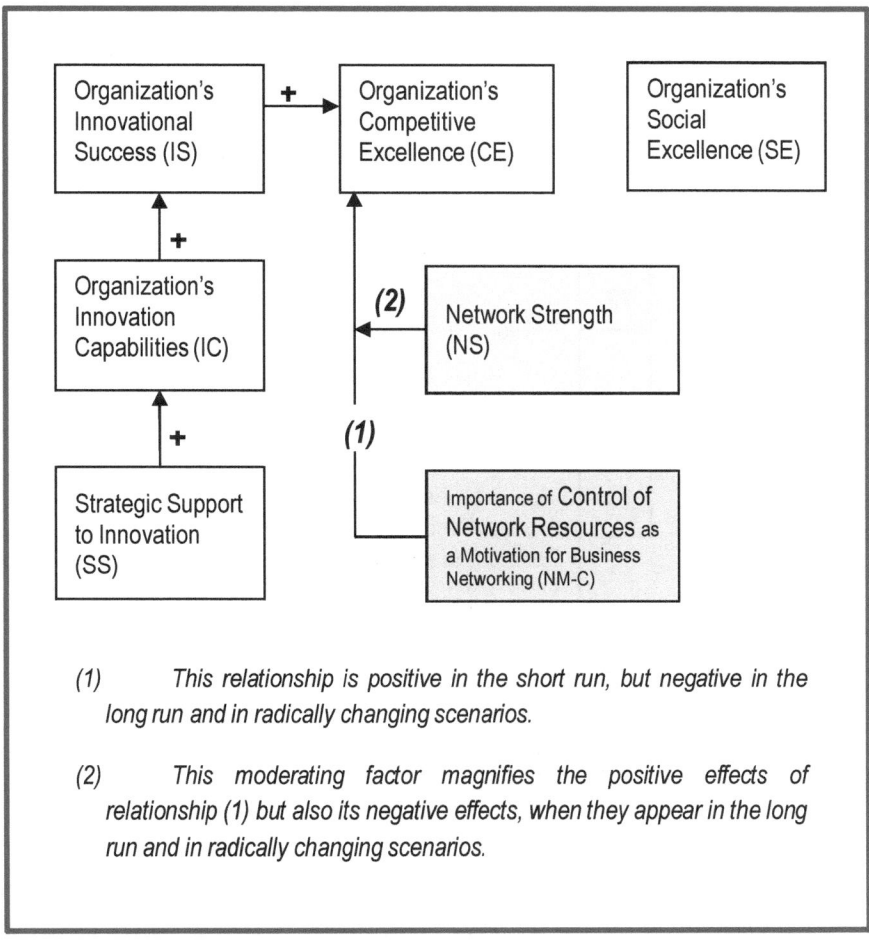

(1)   This relationship is positive in the short run, but negative in the long run and in radically changing scenarios.

(2)   This moderating factor magnifies the positive effects of relationship (1) but also its negative effects, when they appear in the long run and in radically changing scenarios.

Fig. 6.3 Influence of Business Networking (measured through Control of Network resources as a Motivation and through Network Strength) on innovation-related processes and performances

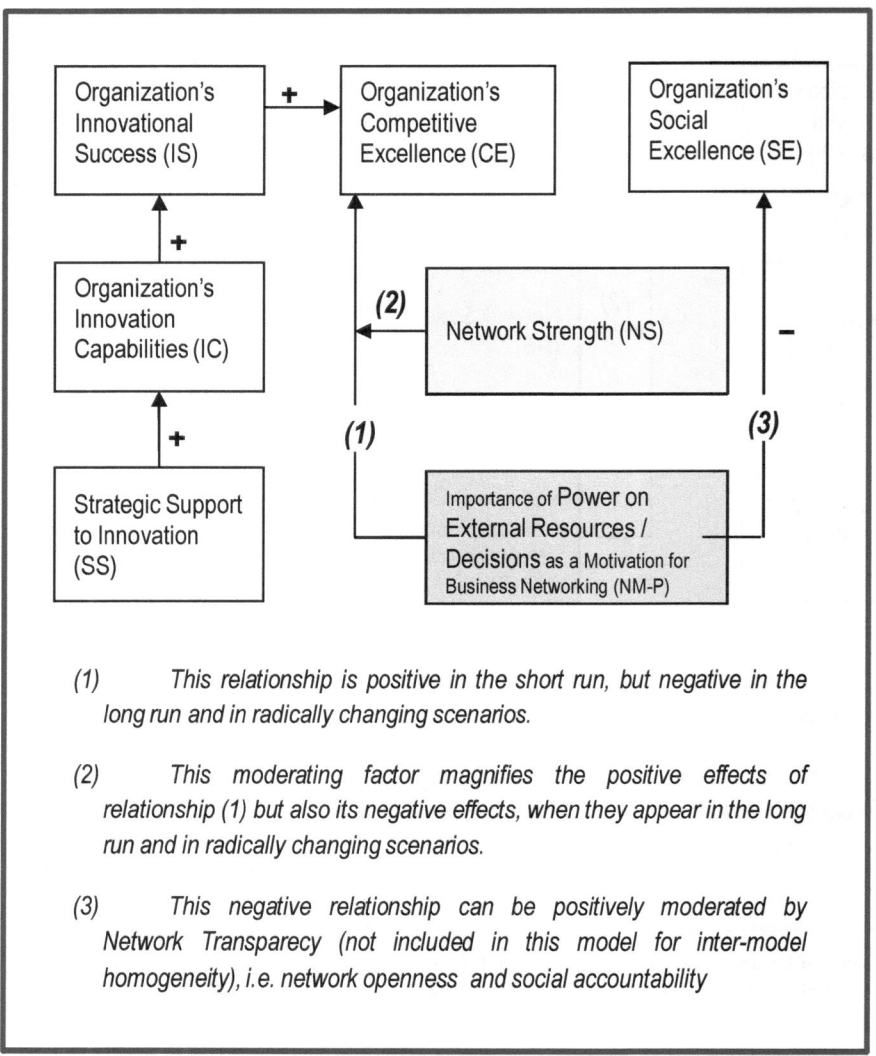

(1)   This relationship is positive in the short run, but negative in the long run and in radically changing scenarios.

(2)   This moderating factor magnifies the positive effects of relationship (1) but also its negative effects, when they appear in the long run and in radically changing scenarios.

(3)   This negative relationship can be positively moderated by Network Transparecy (not included in this model for inter-model homogeneity), i.e. network openness and social accountability

Fig. 6.4 Influence of Business Networking (measured through Power on External resources / Decisions as a Motivation and through Network Strength) on innovation-related processes and performances

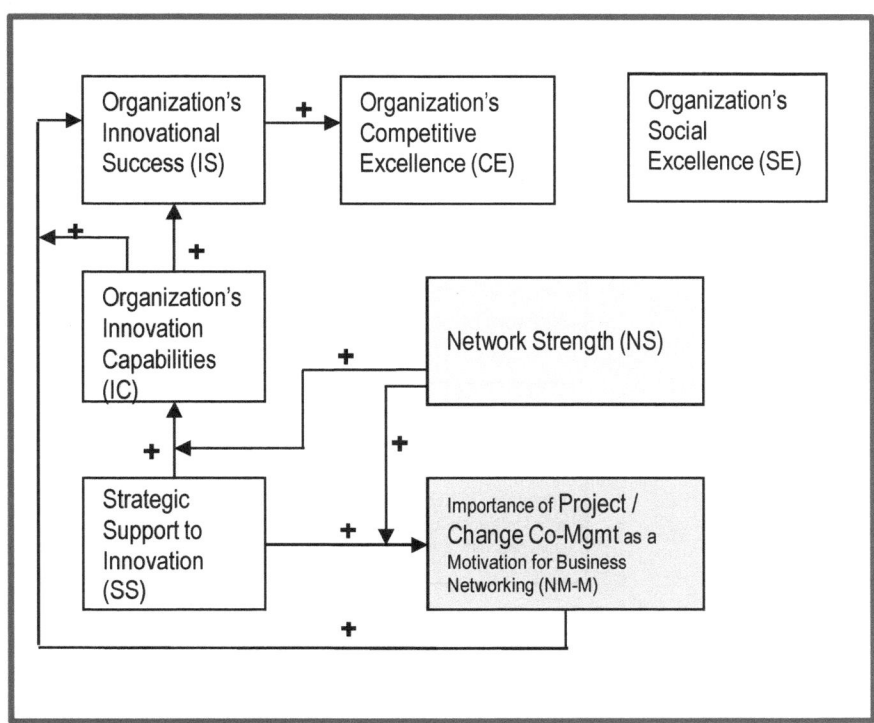

Fig. 6.5 Influence of Business Networking (measured through Project / Change Co-Management as a Motivation and through Network Strength) on innovation-related processes and performances

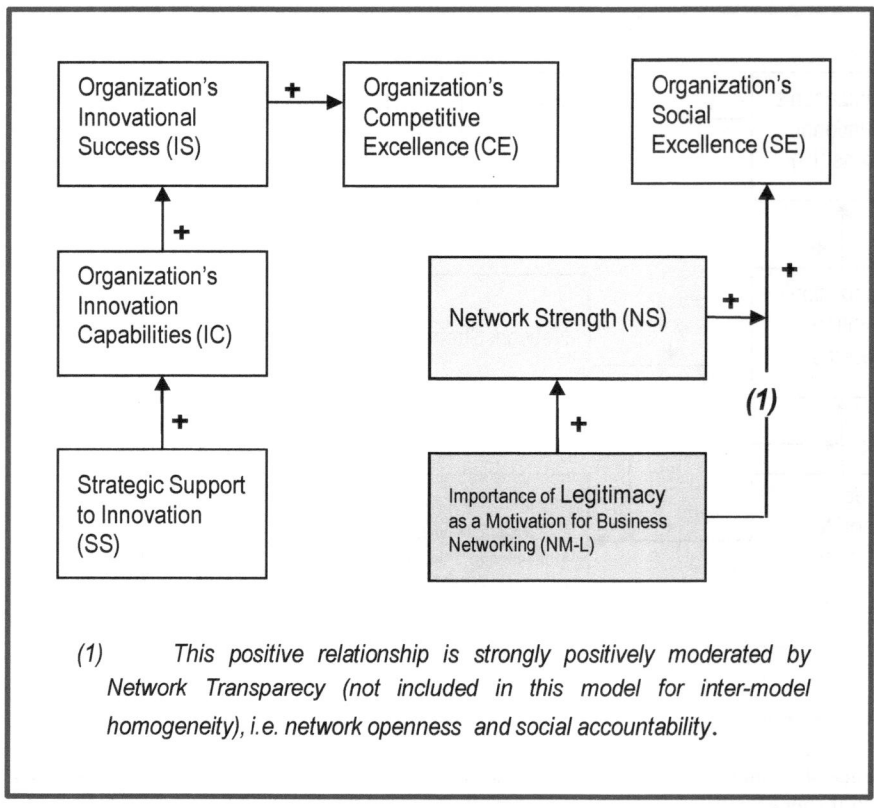

Fig. 6.6 Influence of Business Networking (measured through Legitimacy as a Motivation and through Network Strength) on innovation-related processes and performances

Table 6.3. Scale: Strategic Support to Innovation (developed in Chapter 4)

Strategic Support to Innovation (SS)

Do you agree with the following statements? Please consider your organization's situation in the last five years. (1=strongly disagree, 5=strongly agree, 3=neutral).

SS-1   Our top management wants to position our organization as a unique one in its industry in the way it operates.

SS-2   Our top management is committed to diversify our products/activities and to enter new markets.

SS-3   Our organization strategically aims at pioneering new products/services.

SS-4   Our organization strategically aims at the highest quality and specialization of products/services.

SS-5   In our organization, intensely innovative periods alternate with consolidation periods, in which the achievements are rationalized and efficiency is consolidated.

SS-6   Our management periodically commissions professional market surveys, and/or SWOT-diagnostic studies, reorganization studies, morale surveys, customer satisfaction surveys, etc., to identify new opportunities and areas of innovation and improvement.

SS-7   Our organization has created at least one area, or department, or business unit, or spin-off or joint venture, that is quite protected from short-term cost/revenues pressures and has the mission to develop radically new capabilities or entering completely new markets.

SS-8   Our organization has developed privileged relationships with those external subjects (e.g. suppliers) that may allow us to develop radically new capabilities or entering completely new markets.

Table 6.4. Scale: Innovation Capabilities (developed in Chapter 4)

Innovation Capabilities (IC)

Do you agree with the following statements? Please consider your organization's situation in the last three years. (1=strongly disagree, 5=strongly agree, 3=neutral).

IC-NG	**First Dimension: Capability of Novelty Generation**
IC-NG1	Our managers use benchmarking to generate new and useful ideas.
IC-NG2	Our management invests in the cooperation with customers, suppliers, competitors and/or research institutions for exchanging, securing or testing out innovative ideas.
IC-NG3	Our organization strategically aims at offering personalized products/services to customers.
IC-NG4	In our organization, there is at least one high-level and influential manager coming from other industries or from different professional experiences.
IC-NG5	Our organization sometimes hires people coming from the best competitors.
IC-NG6	In our organization, initiative and experimentation of new solutions are encouraged and rewarded.
IC-NG7	(reverse item) In our organization, complying with habits and procedures is more rewarding than finding new solutions.
IC-NG8	Our organization's managers are very active in exploring the environment (e.g. by establishing new relationships, learning new things, visiting new markets), even if this is not directly aimed at a specific business goal.
IC-NG9	Our organizations' policies include structured initiatives for personnel's education and retraining, and/or encourage the employees' participation in seminars, conferences, professional associations, cultural circles, etc.
IC-NG10	In our organization, there are incentives for employees to point out inefficiencies and problems and to suggest new solutions and improvements.
IC-NG11	(reverse item) In our organization, people who are obsequious with their bosses and unfair with colleagues are more likely to advance their careers.
IC-NG12	In our organization, those who take decisions can easily access reliable and updated information about customers' feedbacks, stakeholders' expectations, suppliers' problems and potentialities, etc.
IC-NG13	Bright, innovative young professionals are recurrently hired and given challenging assignments in our organization.
IC-NG14	Our organization has activated effective cooperation or partnerships with other businesses, research institutions, etc., in order to achieve superior know-how or technological competences.
IC-NM	**Second Dimension: Capability of Novelty Management**
IC-NM1	In our organization, there are several very experienced people, who know our business and our industry very well.
IC-NM2	Our organization is capable of effectively adopting successful solutions developed elsewhere, thanks to systematic benchmarking, updating initiatives, and/or market exploration.
IC-NM3	Our management takes a more sophisticated approach than our competitors', and is more familiar with advanced IT solutions.
IC-NM4	Our organization has proved capable of effective planning and project management.
IC-NM5	Our organization has proved capable of effective auditing.
IC-NM6	Our organization has proved capable of effective process management.
IC-NM7	When an innovation is decided for in our organization (e.g. a new on-line sales channel), a multifunctional or multidisciplinary team or task force (e.g. sales+IT+warehouse managers) is usually established in order to manage the innovation process.
IC-NM8	After being implemented, innovations are systematically and constructively monitored, in order to identify successes, problems and possible areas of improvement.
IC-NM9	Our organization has a strong identity and a recognized tradition as for products/services, quality levels, and way of working.
IC-NM10	Hard work is part of our organization's culture.

Table 6.5. Scale: Innovational Success (developed in Chapter 2)

Innovational Success (IS)

Do you agree with the following statements? (1=strongly disagree, 5=strongly agree, 3=neutral).

IS-1   Our organization has an excellent image of being innovative, and our innovations are often a benchmark for competitors.

IS-2   An important part of our current revenues were derived from recent product / service innovation.

IS-3   Our organization has implemented a stream of successful innovations in business processes and/or organizational culture.

IS-4   Our organization has re-designed its competitive environment, for example by entering new markets or by modifying its supply network.

IS-5   (reverse item) An important part of our organization's innovation projects has recently failed, because it proved incompatible with the status quo (e.g., with old work habits, with established powers, with our legacy information system, etc.)

IS-6   (reverse item) An important part of our organization's innovation projects failed in the last years, because their costs proved too high.

IS-7   (reverse item) Sometimes I fear that we are too specialized or too dependent from specific solutions or specific conditions, and that we would be hardly capable to adapt to changing conditions.

IS-8   Our organization has not made serious mistakes in choosing the technological tools for supporting innovations in the last years.

Table 6.6. Scale: Competitive Excellence (developed in Chapter 2)

Competitive Excellence (CE)

Do you agree with the following statements? (1=strongly disagree, 5=strongly agree, 3=neutral).

CE-1   Our organization's profitability has satisfied our investors/owners.

CE-2   (reverse item) Our organization's sales have performed poorly, if compared to the competitors'.

CE-3   The financial strength of our organization is satisfactory for our investors/owners.

CE-4   Our organization has proved capable of adapting to market challenges.

CE-5   Our organization has displayed superior operating efficiency in comparison with our competitors in the last 3 years.

Table 6.7. Scale: Social Excellence (developed in Chapter 2)

Innovational Success (IS)

Do you agree with the following statements? (1=strongly disagree, 5=strongly agree, 3=neutral).

SE-1   Our organization implements effective programs to minimize its negative impact on the environment.

SE-2   Our organization supports non-governmental organizations working in problematic areas and/or contributes to campaigns and projects that promote the well-being of the society.

SE-3   Our organization effectively contributes to improve the value and image of the territory (e.g., the city) it operates in.

SE-4   Our organization supports schools and/or universities and/or research centers, and/or funds scholarships.

SE-5   Our organization supports employees who want to acquire additional education.

SE-6   The managerial decisions related with the employees are usually fair in our organization.

SE-7   Our organization selects socially responsible business partners (e.g. suppliers) only.

SE-8   Our organization targets interaction fairness in its business relationships.

SE-9   Our organization's image has been damaged by boycotts or scandals or legal actions in the last three years.

SE-10   Our organization complies with legal regulations completely and promptly.

Table 6.8. Scale: Business Network Strength (developed in Chapter 3)

Business Network Strength (NS)

Do you agree with the following statements? (1=strongly disagree, 5=strongly agree, 3=neutral).

NS-IR	**First Dimension: Interaction Repetitiveness**
NS-IR1	Many business relationships within our Business Network are so trustful and cooperative that it would be hard to replace them, both for us and for our Network Partners.
NS-IR2	In many relationships within our Business Network, the reciprocal knowledge between Network Partners is so specific and accurate, that both partners can simplify the reciprocal interaction activities, saving money and energies.
NS-IR3	Within our Business Network, important and effective investments have been made in some relationships, and as a consequence the interruption of such relationships is unlikely.
NS-IR4	Within our Business Network, there are organizations with divergent interests, that in my opinion are likely to quit.
NS-TR	**Second Dimension: Network Transparency**
NS-TR1	During the interactions with our Business Partners, we spend time also telling and commenting on experiences and events occurred to other Business Partners.
NS-TR2	In our Business Network, the news about a wrong or mistake made by anyone in our milieu immediately spread and are made known to all the Network Partners.
NS-SM	**Third Dimension: Capability to Select and Manage Interactions**
NS-SM1	Our organization promotes cooperative and intense communication with those external subjects (e.g. faithful customers, strategic suppliers, banks, associations, etc.) with which recurrent business interactions take place.
NS-SM2	(reverse item) In our organization, it is not clear who is expected to contribute to select, and authorize investments on, the most important long-term business relationships.
NS-SM3	The external subjects with which our organization carries on long-term relevant business relationships are chosen carefully, with the contribution of all the areas involved in each specific relationship (e.g. Sales people, Operations people, Information Systems people).
NS-SM4	Our organization developed special procedures (e.g. better payment conditions, simplified bureaucracy, top managers' direct commitment) to manage the business relationships identified as strategic or long term ones.
NS-SM5	Our organization developed specific procedures or criteria aimed at tuning up the risk levels accepted in a certain business relationship according to its trustworthiness and importance.
NS-SM6	Our organization developed appropriate and effective technological solutions (e.g. information systems, logistic systems) dedicated to optimize long-term business interactions.
NS-SM7	Our organization has proved capable of successfully designing and managing complex contracts for long-term business relationships, such as outsourcing contracts, joint-ventures, etc.
NS-CP	**Fourth Dimension: Capability to Manage Conflicts and to Punish Opportunist Behaviors**
NS-CP1	Within our Business Network, the importance and usefulness of the cooperative relationships between Network Partners are strongly perceived.
NS-CP2	Within our Business Network, there is a reciprocally positive attitude towards the values and culture of the other Network Partners.
NS-CP3	(reverse item) Our Business Network may work better, if only the network Partners had not so different languages and communication styles.
NS-CP4	Should a Business Partner behave unfairly with our organization, we would not take that lying down even if this carries risks and costs.
NS-CP5	In our Business Network, who does not behave is cut out from the whole system, or in any case is blamed and punished by all the Network Partners.
NS-CP6	In our Business Network, some relationships are ruled by special long-term contracts or agreements, such as alliances, joint-ventures, or outsourcing service level agreements.
NS-CP7	In our Business Network, there are reliable mediators and/or authoritative network institutions, such as associations or network leaders, capable of equitably solving conflicts and punishing unfair behaviors.

Table 6.9. Scale: Importance of Reliability/Predictability as a Motivation for Business Networking (developed in Chapter 5)

Importance of Reliability/Predictability as a Motivation for Business Networking (NM-R)

Do you agree with the following statements? (1=strongly disagree, 5=strongly agree, 3=neutral).

NM-R1   In our business context, it is often better to deal with trustful and reliable counterparts, even at the cost of renouncing attractive alternative proposals or possibilities.

NM-R2   A great deal of our organization's success stems from our policy of excluding potentially unreliable subjects from our business relationships, even if they could be potentially interesting.

NM-R3   Our organization's success is strongly rooted in our reputation of being trustful, fair and rule-abiding.

NM-R4   (reverse item) In our milieu, business relationships tend to be very numerous, extemporary and impersonal: resources are invested in finding possible substitutes (e.g. new customers, new suppliers, etc.) more than in the existing relationships.

NM-R5   In our business context, customers are usually very loyal once good product quality is assured.

Table 6.10. Scale: Importance of Information Exchange as a Motivation for Business Networking (developed in Chapter 5)

Importance of Information Exchange as a Motivation for Business Networking (NM-I)

Do you agree with the following statements? (1=strongly disagree, 5=strongly agree, 3=neutral).

NM-I1   In our milieu, a good network of relationships allows access to business-relevant information and advice, which otherwise would be hardly available.

NM-I2   Most of the information allowing true competitive advantage for our organization were accessed through our organization's network of relationships.

NM-I3   Our organization invests to provide managers and employees with occasions (e.g. participation in trade fairs, conferences, associations, clubs) to share valuable information with established or potential business partners.

NM-I4   Our managers are encouraged to leverage our business network to collect valuable information and to make it available for the whole organization.

NM-I5   Our organization supports associations or organizations specifically aimed at business-related sharing of information and knowledge..

NM-I6   Thanks to our long-term business relationships, our organization gained in-depth knowledge of many important actual and prospect business partners, of their behaviors, needs, cultures, structures, markets, etc.

Table 6.11. Scale: Importance of Control of Network Resources as a Motivation for Business Networking (developed in Chapter 5)

Importance of Control of Network Resources as a Motivation for Business Networking (NM-C)

Do you agree with the following statements? (1=strongly disagree, 5=strongly agree, 3=neutral).

NM-C1   In our milieu, the weakest partners in B2B networks (e.g., small suppliers) are often over-exploited and swamped by the strongest organizations (e.g. key clients, or leading dealers/intermediaries).

NM-C2   Our organization is, or will probably be, strongly bonded to another company through acquisition, board interlocks, joint-venture or long-term formal alliance.

NM-C3   In our milieu, important resources must be dedicated to accurately define and legally secure the organization's relationships with the other parties, for example through long-term contracts, detailed service level agreements, etc.

NM-C4   (reverse item)In our milieu, contracts are sometimes considered as just sheets of paper, or as annoying factors of rigidity: trust and good personal relationships are often more effective.

Table 6.12. Scale: Importance of Power on External Resources and Decisions as a Motivation for Business Networking (developed in Chapter 5)

Importance of Power on External Resources and Decisions as a Motivation for Business Networking (NM-P)

Do you agree with the following statements? (1=strongly disagree, 5=strongly agree, 3=neutral).

NM-P1   (reverse item) In our organization's milieu, a well-structured, competitive and innovative organization usually succeeds in business, even without special protections (such as political support or friends in high places).

NM-P2   Political decisions may impact our sector heavily: joining forces also with competitors, to make pressures against damaging decisions, is a good investment.

NM-P3   In our sector, the main role of professional / industrial associations should be protecting the members against system threats such as excessive bureaucracy, excessive taxes, or excessive competition.

NM-P4   In our organization's milieu, a potential business partner (e.g. a supplier, a distributor, a professional firm) accredited as having friends in the high places is often preferred to a more competent and reliable, but not so well-connected subject.

NM-P5   (reverse item) In our organization's milieu, the procedures for getting a positive response from the Public Administration (e.g. authorizations, public funding, subsidies, verdicts) are efficient, transparent and equal for all.

Table 6.13. Scale: Importance of Project / Change Co-Management as a Motivation for Business Networking (developed in Chapter 5)

Importance of Project / Change Co-Management as a Motivation for Business Networking (NM-M)

Do you agree with the following statements? (1=strongly disagree, 5=strongly agree, 3=neutral).

NM-M1   In our industry, many of the most successful organizations have effectively integrated their processes with their main customers and/or suppliers and/or partners, for example by sharing databases, or by co-automatizing order processing.

NM-M2   A great deal of our organization's success stems from innovation projects we carried on in cooperation with other subjects, such as e.g. suppliers, customers, research institutions, Public Administration bodies, banks, etc.

NM-M3   Our organization successfully relies on third parties for critical activities, such as, for example, IT innovation, market analyses or internationalization projects.

NM-M4   (reverse item) In our milieu, the organizations rarely cooperate on specific projects, because the organizations often don't like the idea of sharing confidential processes and information.

NM-M5   (reverse item) In our organization's milieu, the organizations rarely cooperate on specific projects, because people often don't like the idea of the reciprocal dependence that process and/or project integration may result in.

Table 6.14. Scale: Importance of Legitimacy as a Motivation for Business Networking (developed in Chapter 5)

Importance of Legitimacy as a Motivation for Business Networking (NM-L)

Do you agree with the following statements? (1=strongly disagree, 5=strongly agree, 3=neutral).

NM-L1   In our industry, becoming a supplier or a partner of a leading company can boost an organization's reputation even dramatically.

NM-L2   In our industry, there are some recognized best practices for innovation, that most organizations seek to imitate.

NM-L3   If an organization wants to be legitimated as reliable in our sector, it must comply with a complex system of norms and procedures.

NM-L4   (reverse item) In our milieu, smaller or younger organizations tend to adapt to a leading organization's culture, practices and systems, so that their interactions with the leader are smoother and stronger.

6.8 Limitations and Further Research

Literature is growingly inviting managers to concentrate on the importance of business networks in order to enhance their organization's capabilities to innovate and in order to better address market challenges.

It is, then, responsibility of the scholarly community to provide managers and decision-makers with tools for evaluating business networks and network bonds, for choosing them, for modifying them, for assessing opportunities and threats hidden in them.

This book seeks to contribute to this goal, by addressing the "theoretical fragmentation" that affects management and organization studies on business networks.

I sought to utilize several important extant theories and to exploit their complementarities to build a comprehensive theory of business networking, capable to explain the whole range of both positive and negative impacts of business networking on innovation processes and innovation-related performances.

Limitations of this theory-building work can be synthesized as follows:

1. an important theoretical approach, i.e. the social network analysis, has been marginally taken into consideration in the theory-building phase of this study;
2. further and more systematic qualitative field researches would be advisable, in order to corroborate and fine-tune the final models; and
3. the complex phenomena occurring when two or more Motivations for Networking co-should be more thoroughly investigated, preferably by longitudinal researches.

Once the models are fine-tuned and confirmed, quantitative testing could be a fertile field for further research.